T0234774

Mathematics to the Rescue of Democracy

Paolo Serafini

Mathematics to the Rescue of Democracy

What does Voting Mean and How can it be Improved?

 Springer

Paolo Serafini
Department of Mathematics
Computer Science and Physics
University of Udine
Udine, Italy

ISBN 978-3-030-38367-1   ISBN 978-3-030-38368-8   (eBook)
https://doi.org/10.1007/978-3-030-38368-8

This Springer imprint is published by the registered company Springer Nature Switzerland AG
The registered company address is: Gewerbestrasse 11, 6330 Cham, Switzerland

Dedicated to the dear memory of my friend and colleague Bruno Simeone (1946–2010), who, some years ago, introduced me to the mathematics of electoral systems.

Preface

The idea of writing this book came to my mind as a result of the Italian political elections in 2018. The election outcome has produced a host of debates in the media that conveyed the impression of a general confusion of ideas. The complexity, both theoretical and technical, of the voting activity seemed to be totally absent within these debates. This confusion continues, notably not only in Italy, and the voting proposals that periodically enter into the political discussion are based on emotion and whatever issues currently perceived as being important or convenient, never on sound scientific reasoning.

The aim of this book is to provide at least some knowledge to a wider audience on the fundamentals of voting theory in the illusive hope of making more people aware of the very meaning of voting and of the various possibilities for improving the present voting systems. Maybe this could be of some help in dealing with delicate questions that are better answered outside of the realm of partisan points of view.

Pieris, Italy
September 2019

Paolo Serafini

Acknowledgements

A particular thanks is due to my wife Viviana, who encouraged me to write this book on a subject that is new to me, one of 'almost' scientific-popular writing, and who later contributed to improving the drafting of some chapters.

Contents

1	**Introduction**	1
2	**Voting**	3
3	**Aggregating Different Evaluations into One Unique Evaluation**	7
4	**Condorcet**	11
	4.1 Majority Principle	11
	4.2 Condorcet's Principle	13
	4.3 Condorcet Cycles	15
	4.4 Principle of Independence of Irrelevant Alternatives	18
	4.5 Choice Among Many Motions	19
5	**Borda**	21
	5.1 Rankings and Points	21
	5.2 Difficulties in Borda's Method	23
	5.3 Difficulties in Range Voting	26
6	**Simple Majority and Run-Off Voting**	29
	6.1 Problems with Simple Majority	29
	6.2 Run-Off Voting	31
	6.3 Instant Run-Off Voting	32
7	**Impossible Wishes**	35
	7.1 Arrow's Impossibility Theorem	35
	7.2 Single Vote and Arrow	38
	7.3 Condorcet and Arrow	39
	7.4 Approval Voting and Arrow	39
	7.5 Only Two Alternatives	41

7.6   Non-manipulability of an Electoral System   42
7.7   Participation Criterion   44
7.8   Choice and Rank Monotonicity   45

8   Majority Judgement   47
8.1   The Median and the Majority Grade   48
8.2   Grade Language   51
8.3   Collective Ranking   52
8.4   Domination in Majority Judgement   55
8.5   The 2012 French Presidential Elections   56
8.6   Majority Judgement and Condorcet   59
8.7   Majority Judgement and the Impossible Wishes   63
8.8   Possible Future Developments   67

9   Legislative Territorial Representation   69
9.1   General Criteria   69
9.2   Largest Remainder Method   72
9.3   Paradoxes   73
9.4   Divisor Methods   76
9.5   Seats of the European Parliament   81

10   The United States Congress   83

11   Legislative Political Representation   87
11.1   Foreword   87
11.2   Choice of the Candidates   88

12   Biproportional Representation   93
12.1   Biproportional Apportionments   93
12.2   A Simple Case: One Seat per District and Two Parties   99
12.3   The Italian Electoral 'Bug'   100

13   Political Districting   103
13.1   Territory Subdivision   103
13.2   Equity of a Subdivision   105
13.3   Criteria for Fair Political Districting   109

14   One Person–One Vote   113
14.1   Voting Power   113
14.2   Voting Power for Weighted Votes   116
14.3   The U.S. Presidential Election   118

15   Some Additional (and Personal) Considerations   123

References   127

Index   133

About the Author

Paolo Serafini was Professor of Operations Research at the Department of Mathematics, Computer Science and Physics of the University of Udine until 2016, when he retired. He is still participating in academic life as a Professor Senior. In the academic year 1995–96, he was Visiting Professor at Carnegie Mellon University, Pittsburgh, and in 1980, he was Visiting Researcher at the University of California, Berkeley. His scientific interests have dealt mainly with the development of mathematical models for resource optimization and, in recent years, with the mathematics of electoral systems as well.

Chapter 1
Introduction

In 1925, the famous and influential journalist Walter Lippmann wrote in his book The Phantom Public ([59] p. 46): "What in fact is an election? We call it an expression of the popular will. But is it? We go into a polling booth and mark a cross on a piece of paper for one of two, or perhaps three or four names. Have we expressed our thoughts on the public policy of the United States? Presumably we have a number of thoughts on this and that with many buts and ifs and ors. Surely the cross on a piece of paper does not express them. It would take us hours to express our thoughts, and calling a vote the expression of our mind is an empty fiction."

Lippmann was right. The present way of voting is rough and may produce electoral outcomes that are puzzling either because they don't seem to reflect the wishes of the majority of voters or because they lead to deadlock situations.

Voting is a fundamental activity in the life of a society, and it would be appropriate to know exactly what kind of outcome we want from an election, what the theoretical limitations of voting are, what kind of information would be helpful for the electors to pass along and, finally, how the efficacy of voting could be improved.

In general, people have some minimum knowledge about medicine, economics or other disciplines, and this enables them, at least partially, to direct their actions. Instead, in the practice of voting, the impression is that there is very little to understand, because everything seems obvious and elementary. On the contrary, it is much more complex than it appears.

The aim of this short book is to explain, in a hopefully simple way, the foundations upon which electoral techniques are based. Although one may think that voting pertains social or juridical sciences, the fact is that the basis of every electoral system is mathematics. What's more, it is not a particularly abstruse form of mathematics. It can be made comprehensible to non-mathematicians. But neither is it elementary. In any case, the effort to understand it can garner the significant reward of a greater awareness of what we do when we vote. The title of the book highlights this hope.

The aim is not to provide a detailed description of certain electoral systems, either in Italy or elsewhere. There will, however, be more than a few references to existing or past situations, either as examples or to underline problems and defects.

© Springer Nature Switzerland AG 2020
P. Serafini, *Mathematics to the Rescue of Democracy*,
https://doi.org/10.1007/978-3-030-38368-8_1

It is important to know that, no matter how an electoral system has been devised, there are goals that are impossible to pursue, independent of the presumptions of the system's authors, but there are also opportunities toward improving the situation in an informed way.

It may be surprising to see that many electoral systems that are currently employed in various countries have evident defects and, yet are, nonetheless, still in use, justified for reasons of habit and familiarity with a particular method. For instance, one of the most fallacious ideas is that the alternative that receives most votes should be chosen even in cases in which it has received less than half of the total votes. Yet, many electoral systems are based on this idea.

A quite common bit of confusion is related to the aim of voting. Voting for parliamentary representation is very different from voting for a president or a govern. If these two aims overlap, the outcomes can be problematic.

The vote that we are accustomed to is the simple indication of a preference for one alternative among many. What we think of the alternatives that we have not chosen is not transmitted as information, and this is unfortunate, because the vote outcome would be in far greater compliance with the collective will than the current method of voting. Is it conceivable to modify a secular habit? However difficult this notion may seem, the time has probably come to make the vote more informative. In this book, we will try to illustrate the advantages of voting in a more completely engaged way. To this purpose, we will make explicit reference in Chap. 8 to the method devised by Balinski and Laraki [12, 15], the so-called *Majority judgement*.

The argument is a very large one and, in order to be thorough, four times as many pages would be necessary. Hence, we have opted to focus on a few select principles in order to make the fundamental aspects understandable and possibly direct the reader to more specific essays for an in-depth study.

Chapter 2
Voting

In the idea of the State that we inherited from the Age of Enlightenment, there are three public powers that belong to the citizens who constitute the State: the legislative power, the executive power and the judicial power. Clearly, there are other powers in the society that can condition, for better or for worse, the life of the State, but it is the three aforementioned ones that belong to the community of citizens. The practice of these powers cannot be carried out by the community as a whole, and thus it is necessary to delegate these tasks.[1]

If we want the three powers to be independent of each other, it is important to understand how the machinery of mandate allows for such independence. There can be differences among nations. For instance, the Italian Constitution guarantees the independence of the judicial power from the other two in a precise way. The mandate goes through a selection that is an expression of the citizens inasmuch it is guaranteed by public laws. It is not the expression of a direct vote. If it were so, it would follow the same logic of the mandate for the legislative and executive powers, thereby jeopardizing the very concept of independence. This separation of the judicial power from the other two is not as sharp in every nation and the very fact that there are often attempts to weaken the separation shows why it is so necessary.

The question is different for the legislative and executive powers. The mandate is direct and is carried out through a vote. The choice of how to implement the mandate changes according to the nation. Basically, there are three possibilities: the vote is for the legislative power and the mandate for the executive power is left to the legislative power, or the vote is for the executive power and the mandate for the legislative power is left to the executive power, or there are two distinct votes, one for the legislative power and another for the executive power.

---

[1]Concerning the idea of 'direct democracy', without mandate and intermediate bodies, see the considerations in the last Chap. 15.

© Springer Nature Switzerland AG 2020
P. Serafini, *Mathematics to the Rescue of Democracy*,
https://doi.org/10.1007/978-3-030-38368-8_2

The first method is the one used in so-called parliamentary democracies: a parliament is elected and the executive power is the expression of the will of the parliament. The second method is not formally present in any democratic country. However, it can, in practical terms, happen. The recent history of the Italian parliament shows that an inversion of the two roles has been attempted. The third method is in effect in a number of important countries, like France and the United States of America.

Having an 'effective' mandate for both the legislative power and the executive power at the same time is problematic, because the legislative power is well delegated when the society's many articulated requests are being addressed, whereas the executive power is well delegated if the government is granted a certain degree of stability, at least for a set period of time. It is not possible to govern well if the government's actions are frequently interrupted or weakened.

Representation and governmental stability are therefore the two opposing requirements of a parliamentary democracy. If the method of voting favors representation, the governing becomes less stable and, vice versa, if governmental stability is favored, the very essence of the legislative power is undermined. In designing a voting method, it must be clear which goal we want to pursue, within the bounds of the constitutional law.

The Italian Constitution has been designed to give ample representation to the citizens, after two decades of dictatorship, with the idea in mind that the parliament would be able to form stable governments. Indeed, this is what happened from 1948 (when the Constitution was approved) until the beginning of the nineties. A fact that has often been misunderstood concerns the large numbers of different governments that have been formed throughout these years. In most cases, each change of government consisted in a turnover of power positions in the administration. The political course of action, contrastingly, was always stable. Things began to change starting with the 1994 elections, when the electoral mechanism was forced to address the choice of the prime minister as well, in spite of the fact that the Constitution had not been changed. This created enormous confusion that has still not dispelled. The idea has been spread that political elections are meant for choosing governments directly. It has been presumed that the vote percentage ranking could constitute the basis for obtaining a government mandate,[2] and there has been a lot of misguided talk about 'winners' and 'losers', terms that do not exist in the Constitution.

The mechanisms for forming a government in a parliamentarian democracy are not based on rankings. The scenario involves parties that have been apportioned a certain number of seats and the goal is to form coalitions that can gather at least half of the total number of seats (unless a single party gets the absolute majority of seats, a very unlikely possibility for many countries, let alone Italy). How this can happen is more pertinent to Game Theory rather than it is to Social Choice Theory. For this purpose, particular indices, like the Shapley–Shubik value [91, 92], have been proposed. These indices evaluate the strength of the parties toward the formation

---

[2]On the danger of considering relative majorities as being eligible to form governments, see Chap. 6.

of a coalition. It should not be surprising that three parties that have respective seat strengths of 45, 35 and 20% actually have the same bargaining power, because they are all equally necessary to obtain the absolute majority of seats.

All of these aspects should be clear in the choice of an electoral mechanism and in the consequent expectation within the population. If the need to entrust the executive power to someone by a direct election is felt to be mandatory, then it seems unavoidable that two separate elections will be necessary, one for the parliament and the other one for the government (in Italy, this would entail a modification of the Constitution). The United States of America have been governed in this way for more than two centuries.

By having separate voting for the legislative and executive powers, there is the risk that the political colors of the two powers will be different, thus creating a disharmony between government and parliament. When this has, indeed, happened, like, for instance, in France with the so-called 'cohabitation', or in the United States with a President at odds with a hostile Congress, no insurmountable problems have occurred. Rather, one gets the impression of a healthy democratic dialectic. It seems more problematic to have a unique vote and compel the parliament to form difficult coalition agreements. The recent examples of Germany, Italy, Spain and Israel demonstrate this fact.

In the subsequent chapters, we shall examine the problems that arise with the choice mechanism, by dealing separately with the questions concerning the formation of a government by electoral methods (Chaps. 3, 4, 5, 6, 7 and 8) and those pertinent to having parliamentary representation (Chaps. 9, 10, 11, 12, 13). We will not deal with the issues connected to the formation of coalitions in parliaments. In Chap. 14, we examine a common feature of both aspects, namely, measuring the voting power of a single voter.

As for the first issue, that is, the choice of a government, the problem of ranking all present alternatives is strictly related. This wider issue is not very relevant in a political context, but it is important in sport, in which ranking is essential. As we shall see in the subsequent chapters, from a mathematical point of view, declaring the winner of a political contest is the same as declaring the winner of some sport competitions.

An issue that must be considered in the future concerns the type of information that electors express by voting. Presently, they express only a preference for one particular alternative, with nothing been said about all of other alternatives. This is clearly very simple and easy to be carried out by all citizens. However, the missing information on the alternatives may cause paradoxical consequences.

A choice method that avoids many of the paradoxes and impossibilities present nowadays in the voting systems, even in more detailed voting methods, is the already cited Majority judgement [12, 15], which might represent an extremely interesting electoral method for the future.[3] In Chap. 8, we shall present the method in great detail by showing how it works and listing its limitations. This method is currently applied in small contests, like sport competitions, wine contests, in which only a few judges have to express a vote, or, put better, a judgement. Transferring this method into a political competition with millions of voting people, might be not so simple. Perhaps, it is only a problem of 'adjustment'. Also, changing an electoral system every few years (like in Italy) creates adjusting difficulties.

The considerations that can be drawn for an electoral system nationwide also hold, on a smaller scale, for choices made at local level. However, even if the same pattern of executive and legislative power is replicated at the local level (the judicial power, at least in Italy, is not a competence of local authorities), one should note that the legislative scope of a local assembly is necessarily quite reduced, whereas the executive scope is not. Therefore, in order to select a local electoral system, the considerations related to a choice of executive type become predominant.

Very often, proposals are brought forth in Italy to write an electoral law, similar to the one for mayoral elections mandating that the name of the future Prime Minister be known as soon as the polls are closed.[4] This would mean completely forgetting about the need for legislative representation at the national level. The national level is different from the local level.

---

[3]It is perhaps useful to read the comments on Majority judgement by three Nobel prize winners in Economics, as quoted in [1]: K.J. Arrow (Nobel prize 1972): 'The authors have proposed a very interesting voting method to remedy the well-known defects in standard methods, such as plurality voting. It requires the voters to express their preferences in a simple and easily comprehensible way, and the authors supply evidence that the candidate chosen by their methods is a reasonable selection. This work may well lead to a useful transformation in election practice'. R. Aumann (Nobel prize 2005): 'Michel Balinski has done it again! He has produced—this time with Rida Laraki—a beautiful, comprehensive, conceptually deep, and utterly sound treatise on the mechanics of democracy. By no means an abstruse, ivory-tower exercise in pure math, the work is supported by a plethora of in-depth empirical analyses taken from real life. Most important, the book introduces a vital new idea that promises to revolutionize democratic decision making: 'judging' rather than voting. Enjoy—while learning'. E.S. Marskin (Nobel prize 2007): 'Balinski and Laraki propose an intriguing new voting method for political elections: have voters 'grade' the candidates and elect the one having the highest median grade. The method will be controversial but deserves careful consideration.'

[4]Mario Monti has dubbed this need as 'Sunday night fever'.

Chapter 3
Aggregating Different Evaluations into One Unique Evaluation

When we deal with the choice of a president or a government through an election, the problem that we have to face is fundamentally that of determining a unique collective evaluation that represents, in the best possible way, an aggregation of the individual evaluations expressed by the voters. We can see that this problem is by no means easy. Moreover, aggregating methods that seem appealing and easily understandable may be subject to serious malfunctions, as we shall see.

If we want to formalize the aggregation problem, we have to introduce 'candidates' (or 'alternatives', as we shall also refer to them) and 'electors' (or 'voters' or 'judges'). Electors express their individual evaluations on the candidates. Starting from all individual evaluations, we have to build up a mechanism that produces a unique collective evaluation on all of the candidates.

We have to specify what type of collective evaluation we want to produce and what type of individual evaluation we want to input into the mechanism. According to the goal that we are pursuing, we have available different alternative ways to define both the collective evaluation and the individual evaluation. The mechanism that aggregates the individual evaluations into the collective evaluation clearly depends on how both evaluations have been defined.

The simplest collective evaluation consists in the choice of a particular candidate (the 'winner'). In this case, there is no collective evaluation of all of the other candidates. However, as a more complex case, we may want to obtain a collective evaluation on all candidates. Indeed, it is very often useful, if not necessary, to have a ranking in order to discriminate among the non-winners. In this case, the collective evaluation is a ranking that specifies who is first, second, and so on. In sport competitions, ranking the athletes is normal procedure. Also, in the choice of a candidate for a job position, it is useful to have a ranking to replace the winner, in case he/she becomes unavailable.

It is also possible to add a further element to the collective evaluation. Beside the ranking, namely, the simple ordering of the candidates, we may add numbers that indicate the 'distances' among the candidates. Hence, we may ask for a collective evaluation that is more and more informative.

© Springer Nature Switzerland AG 2020
P. Serafini, *Mathematics to the Rescue of Democracy*,
https://doi.org/10.1007/978-3-030-38368-8_3

According to the collective evaluation that we want to produce, we have to decide which type of individual evaluation each elector should express. Also, in this case, the individual evaluation may consist of a simple preference for one candidate, or in the indication of a list of preferred candidates, or in a ranking of all candidates, or, even further, in a ranking with numerical indications. These types of evaluations are present in almost all existing methods. In contrast, the already quoted Majority judgement method employs judgements on the candidates, and not rankings. We will discuss this method in detail in Chap. 8.

In this chapter and in the ones that follow we shall try to give an idea of the issues that arise when we have to aggregate the individual evaluations and how these problems can be faced and possibly solved. However, before going into the subject of political choices, it is perhaps useful to look at how the aggregation problem occurs in quite different contexts. Even if the contexts may seem different, their formalization is nonetheless the same, and the methods for solving the problem are therefore the same as well. Furthermore, looking at the same problem in different contexts may cast new light on the context that we are more interested in, i.e., political elections.

As a first example, imagine a group of friends who want to go on vacation. They have in mind different ideas and different preferences, but one thing is sure: they want to go on vacation together. How to choose the place to go? Very likely, they will have a discussion and exchange opinions, trying to influence each other. But suppose that, nonetheless, they do not reach a common decision, so that they eventually decide to solve the question 'democratically' by voting. At the moment, we will not give any clue as to how they will choose to do it. In the following chapters, we will discuss on the various possibilities at length. After all, there is no formal difference between the election of a president and the decision about where to go on vacation: there are individual evaluations out of which a collective evaluation must be extracted. For the time being, readers may feel free to figure out a solution to the vacation problem themselves and then compare it with the ideas that will be explained later.

In considering more realistic cases, sport competitions offer many examples. In several competitions, that which has to be solved is exactly the pattern just shown. In diving, artistic gymnastics and figure skating, the final ranking among the athletes (the 'candidates') is computed starting from the rankings produced by a group of judges (the 'electors'). Usually, each judge assigns a numerical vote to each athlete and de facto produces a ranking at the end of the race. The final score is simply computed by taking the average of the individual scores. Sometimes, this scheme is modified by dropping the best and the worst evaluations from the score of each candidate. In some cases (for instance, in figure skating), the judges' scores are not summed, but are rather used to produce individual rankings for the competitors, and it is from these individual rankings that the final collective ranking is obtained (see the discussion in [15]). But, however sensible these methods may seem, they can raise paradoxical results.

Formula One World Championship is also a significant example: one must aggregate the results of the individual races to obtain the final score of the World Championship. Presently (2019), in each race, points are assigned to the first ten drivers, with one additional point being assigned for the fastest lap, after which one simply adds up the points of each driver. For this example, each single race is like an elector that expresses its own numerical ranking, with the candidates obviously being the drivers. The scoring system in Formula One has, to date, been changed 29 times since 1950, a clear indication that any method has its unsatisfactory aspects. A peculiarity of some past scoring systems was that, for each driver, the sum of the points was limited to a fixed number of races, clearly to the most successful ones. Similar considerations can be drawn for motorcycle races.

The use of points to build up the general score can also be found in the Tours de France cyclic competition from 1905 until 1912 (the race started in 1903 with the time score system) and in the first Giri d'Italia (Tours of Italy) from 1909 until 1913. Since 1913 in France and 1914 in Italy onwards, the time score has been employed, and this is now the traditional method for all cycling races held in stages. The mechanism of summing up the timings of the individual stages to produce the final score seems perfect. After all, it is as if the single stages were merged into one very long unique race. Apart from the drawback that one has to drop out of the race if a stage cannot be completed (thus it would be impossible to apply this method to Formula One), one can always argue that thirty seconds of delay in a crono stage have a different meaning than the same seconds in a mountain stage.

If the 'electors' are not homogeneous, the aggregation presents a further complication. The decathlon example in athletics is significant. As we know the athletes (the candidates) have to take part in ten different track and field competitions (the electors) that include racing, throwing and jumping. For each competition, the result is transformed into a score, and eventually the scores are summed up. Over the years, the scoring has been changed many times and, presently, a complex formula is employed.[1] From the formula, the same increment in a performance is worth a larger score if it is obtained at a higher performance level. For instance, the difference between $10''3$ and $10''2$ for 100 m is worth 24 points, whereas between $11''3$ and $11''2$ is worth 22 points.

Wine competitions have existed since ancient times[2] and they are particularly important nowadays due the economic impact that they have produced. In these contests, several experts (the 'electors') taste the wines (the 'candidates') and assign scores. The mechanism to producing the final score may vary from competition to competition. In a famous competition in 1976 between four French wines and six Californian wines (see [14]), eleven judges had to assign a score from 0 to 20, and the

---

[1] For racing, the formula is $P = a \cdot (b - T)^c$, where P is the score and T is the running time of the athlete; for jumping and throwing, it is $P = a \cdot (M - b)^c$, with M being the jump or throw measure. The parameters a, b and c are defined for each competition. For instance, for the 100 m race, $b = 18$, i.e., running in 18 s corresponds to a score of zero. Apparently, the parameter b represents the worst conceivable performance for an athlete (for the pole vault, it has been set to one meter!). The parameter c varies from a minimum of 1.05 (shot put) to a maximum of 1.92 (110 m hurdles). The parameter a has the purpose of normalizing the different scores so that 1000 points correspond to a very good performance for a decathlete.

[2] Plinius the Eldest, in his Historia Naturalis, Liber XIV, par. 61 et seq., so ranks the wines according to the Emperor's taste: Divus Augustus Setinum praetulit cunctis ... Secunda nobilitas Falerno agro erat ... At tertiam palmam varie Albana urbi vicina... Quartum curriculum publicis epulis optinuere a Divo Iulio ... Mamertina circa Messanam in Sicilia genita.

final score was simply obtained as a sum (without excluding the worst and the best evaluations). The shocking result for the French was that the winner was a Californian wine (although the four French wines were classified at the second, third, fourth and fifth places). In the detailed discussion in [14], all of the paradoxes of this scoring method are put into evidence.

One more example can be found in music competitions in which a jury has to formulate a ranking among a certain number of performers. In this case, there can be eliminating rounds to reduce the number of competitors to a few good performers. Each competition has its own scoring method (in several cases, the first prize may not be assigned, if none of the performers have been judged to have earned it). Many examples can be quoted in which the final outcome has generated harsh resentment among the judges, not only because an evaluation is necessarily subjective, but also because it may happen sometimes that the final ranking, due to its inner mechanism, which may generate paradoxes, produces a result that nobody can acknowledge (see the example on Sect. 7.6). If the judges, as generally happens, do have favorite performers among the candidates, the voting becomes 'strategic', i.e., a judge votes not only in favor of his/her preferred candidate winning but also in service of the other candidates losing. The presence of a strategic will may alter the score significantly and produce results that are surprisingly different from the ones one has in mind.

After this short presentation of competitive contexts of non-political nature, in the following chapters, we will deal only with the problem of electoral choices. However, the reader has to keep in mind that the mathematical 'form' in all of these examples is always the same, so that whatever will be presented hereafter can also be applied to the competitive contexts that we have shown (and even to the vacation example).

Chapter 4
Condorcet

We may say that the modern and scientific approach to the problem of Social Choice began with the *Essai sur l'application de l'analyse à la probabilité des décisions rendues à la pluralité de voix* (1785) by Condorcet [30].[1]

The method proposed by Condorcet has actually already been devised by Ramon Lull [48] some centuries earlier, at the end of 1200. Lull's writings, however, remained unknown up to mid-1900. Only recently have they become the object of an in-depth study. However, it seems certain that Condorcet was not aware of Lull's ideas. One can see the detailed presentation in [95].

In the quoted essay, Condorcet addresses the problem of how an assembly might be able to choose a single alternative among a set of different alternatives that have been formulated and proposed for a vote.

4.1 Majority Principle

Before going into a detailed presentation of Condorcet's approach, let us suppose that voters can only express their preferred alternative, exactly as it happens in the present voting systems. This way of voting is called *Single vote*. If a particular alternative is preferred by the majority of voters, that is, it receives the majority of single votes, then it seems natural that the collective choice is this alternative.

By majority, we mean that more than half of the total number of voters opted for that choice, not simply that it received the most votes out of all of the options.

---

[1] His full name was Marie Jean Antoine Nicolas de Caritat, Marquis de Condorcet (1743–1794). Despite being part of the aristocracy, he participated enthusiastically in the French Revolution and was among the group that drafted the constitution. He believed that the final text of said document had betrayed the revolution's initial intentions and complained about it, perhaps too vociferously for that time, as he ended up having to go into hiding. After some time he emerged out of his hiding place, disguising himself by wearing common clothes, but his polished way of speaking betrayed him and he was caught and imprisoned. The morning after his incarceration, he was found dead in his cell. The true circumstances of his death were never disclosed.

© Springer Nature Switzerland AG 2020

P. Serafini, *Mathematics to the Rescue of Democracy*,
https://doi.org/10.1007/978-3-030-38368-8_4

In the latter case, we would be talking about *simple* or *relative majority*, which we will discuss later in Chap. 6. To distinguish the first from the second case, it is also common to use the term *absolute majority*, but we will continue to use the term majority here to indicate the absolute majority.

Do we need to justify the majority principle? The answer seems to be no, since the sensibility of choosing what is preferred by a majority over that which is preferred by a minority appears obvious. However, there are examples from the past when the majority of people had opinions that we nowadays consider wrong and unfair. For instance, there have been times in history and places on earth where a majority of people have been in favor of slavery.

Condorcet's point of view in regard to accepting the majority principle is quite pragmatic and well reveals the mentality of the Age of Enlightenment: since whoever governs has to impose rules, it is good, for the social peace, that these rules harm the least number of people, and therefore whoever governs must be an expression of the will of the majority.

However, this motivation is not sufficient to dispel doubts about any given majority's ability to act in a manner that can necessarily be considered 'well'. Hence, we have to face the problem of how ample this majority needs to be for a 'wrong' choice to be quite unlikely. The quoted essay also deals with this issue and, indeed, the term 'probability' that appears in the title refers to this type of considerations. It is exactly because of the need to have a low probability of error regarding fundamental decisions, that voting systems addressing these issues often require a majority much larger than half.

"The very essence of democratic government consists in the absolute sovereignty of the majority; for there is nothing in democratic States which is capable of resisting it".[2] So wrote Alexis de Tocqueville almost two centuries ago to stress the strong link between democracy and the concept of the majority ([33], Book 1, Chap. 15).

It is also worth once again quoting Walter Lippmann with regard to the majority principle ([59] p. 48): "Yet the inherent absurdity of making virtue and wisdom dependent on 51 per cent of any collection of men has always been apparent. The practical realization that the claim was absurd has resulted in a whole code of civil rights to protect minorities and in all sorts of elaborate methods of subsidizing the arts and sciences and other human interests so they might be independent of the operation of majority rule. The justification of majority rule in politics is not to be found in its ethical superiority. It is to be found in the sheer necessity of finding a place in civilized society for the force which resides in the weight of numbers."

Without neglecting this type of concerns, it does, however, seem legitimate to establish the following criterion:

---

[2]Il est de l'essence même des gouvernements démocratiques que l'empire de la majorité y soit absolu; car en dehors de la majorité, dans les démocraties, il n'y a rien qui résiste. (Book 2, Chap. 7 in the original version).

Majority criterion: *If an alternative is preferred by more than half of the voters, then it must be selected.*

We will see later (Sect. 7.5) that the Majority criterion, according to May's theorem, enjoys some important properties.

4.2 Condorcet's Principle

Of course, it is quite normal for no alternative to satisfy the Majority criterion. It is here that Condorcet's proposal steps in with great originality. The first important aspect consists in the type of information that it is required from the electors. Electors must not limit themselves to indicating their preferred alternative, but must rank all alternatives.

We can express the ranking in two different ways. We may ask each elector what is their most preferred alternative, what is their second preferred alternative, and so on, up to the last alternative, with possible ties between certain choices.

Given two alternatives A and B, we denote by $A \to B$ the fact that a generic elector prefers A to B and, vice versa, by $B \to A$ that B is preferred to A, whereas the indifference between A and B is denoted by $A \sim B$.[3] For instance, if there are four alternatives A, B, C and D, an elector could give the ranking $B \to C \to A \to D$. This ranking also implies that $B \to A$, $B \to D$ and $C \to D$, due to the transitivity principle (i.e., if B is preferred to C and C is preferred to A, it is natural that B is preferred to A).

Another elector could give a different ranking, like $A \to D \sim C \to B$, which also implies $A \to C$, $A \to B$ and $D \to B$. The ranking of a third elector could be $A \to D \sim B \sim C$, which also implies $A \to B$, $A \to C$ and $D \sim C$. The method does not assume any limitation on the rankings that electors can express. In other words, the method must work with any combination of rankings given by the electors.

Alternatively, we may ask each elector to express a preference for each pair of alternatives, by excluding 'non-rational' preferences. We say that a person does not express a rational preference if he/she evaluates three alternatives pairwise as $A \to B$, $B \to C$ and $C \to A$. Due to the transitivity principle, this would imply the cycle $A \to B \to C \to A$, which amounts to saying that A is preferred to A! But this is a contradiction, and therefore we must exclude non-rational preferences.

Also, an evaluation like $A \sim B$, $B \sim C$ and $A \to C$ is considered non-rational because $A \sim B$ and $B \sim C$ imply $A \sim C$, contradicting the evaluation $A \to C$. Analogously, $A \to B$, $B \to C$ and $A \sim C$ is non-rational because $A \to B$ and $B \to C$ imply $A \to C$, contradicting the evaluation $A \sim C$.

---

[3]We use a notation here that is different from the usual mathematical notation that denotes the preference as $A \succ B$ instead of $A \to B$. We feel that the latter notation has a more immediate appeal to a reader who is unaccustomed to mathematics. $A \sim B$ is the standard notation in mathematics to denote the indifference.

From all pairwise evaluations, it is easy to get a ranking if the preferences are, indeed, rational.

The idea that an individual cannot express cyclic preferences is more subtle than it may appear. If the alternatives under consideration can be evaluated according to several criteria, it may happen that, depending on the criterion, the alternatives are preferred in different ways. For instance, if the alternatives A, B and C are three car models and are judged according to the price, the ranking could be $A \rightarrow B \rightarrow C$, whereas, in regard to appearance, the ranking could be $B \rightarrow C \rightarrow A$, and for consumption, it could be $C \rightarrow A \rightarrow B$. Two rankings alone have already produced a cycle. In any case, it is assumed that, eventually, an individual faced with several alternatives is always able to express a rational set of preferences, apparently by merging all criteria into a meta-criterion.

The second particularly important aspect of Condorcet's method consists in the qualitative type of information. One alternative is only said to be preferred to another one, but it is not said by how much. It may seem an oversimplification not to express the extent of this preference as well. In fact, the introduction of quantitative information presents very critical aspects that we will discuss later.

After the formulations of all rankings by the electors, all of the alternatives are pairwise compared and, for each pair of alternatives, one can see which one is preferred by the majority of voters. So, if a majority of electors prefers A to B, we denote this fact as $A \Rightarrow B$ to indicate that, collectively, A is preferred to B. Of course, it may happen that the number of electors that prefer A to B is the same as those that prefer B to A, so that we would have the collective evaluation $A \approx B$.[4] This is unlikely to happen with a large number of electors, but, with a small number, for example, in a jury for a contest, this situation may happen, unless the number of electors is odd (and, wisely, juries are often made up of an odd number of people).

We are now in the position to introduce another criterion to establish which alternative has to be selected:

Condorcet's criterion: *If an alternative is collectively preferred to any other alternative, then it must be selected.*

Condorcet's criterion thus identifies a winner by applying a Majority criterion many times over: if a majority of electors prefers A to B, and another majority (possibly different) of electors prefers A to C, and so on for all other alternatives, then A must be the final choice. Such an alternative is called a *Condorcet winner*.

Condorcet's criterion is coherent with the Majority criterion, in the sense that a choice based on the Majority criterion also satisfies Condorcet's criterion. Indeed, if a choice is preferred by a majority of voters, this majority also expresses a preference for this choice with respect to all others.

It is useful to give some examples to understand how the method works. Assume that there are five electors that express the following rankings on three candidates A, B and C:

---

[4]We have employed a notation that highlights whether we are dealing with an individual or a collective preference. Individually, we denote $A \rightarrow B$ or $A \sim B$, whereas, collectively, we denote $A \Rightarrow B$ or $A \approx B$.

$$1 \text{ elector:} \quad A \to B \to C$$
$$1 \text{ elector:} \quad A \to C \to B$$
$$2 \text{ electors:} \quad C \to A \to B$$
$$1 \text{ elector:} \quad B \to A \to C$$

As can be seen, no candidate satisfies the Majority criterion. Both A and C are the first choice for two electors out of five. With Single vote, they would obtain 40% of the votes, not enough for the Majority criterion. Hence, no winner is immediately available. So, we apply Condorcet's method and, by carrying out all possible comparisons, we get the following table, which we call *Condorcet table*, in which each number indicates the number of times that the alternative corresponding to that row is preferred to the alternative corresponding to that column. The winning values are denoted in boldface:

	A	B	C
A		**4**	**3**
B	1		2
C	2	**3**	

In this example, A outnumbers both B and C, and so A is the Condorcet winner. With Single vote, we are not able to make a choice between A and C, but having more information, that is to say, all rankings among the alternatives, allows us to go deeper into the decision and make the final choice.

The method produces a winner, and, by proceeding recursively, we may obtain the full ranking. If we drop the row and the column corresponding to the winner from the table, we may simply repeat the procedure on the remaining alternatives. We must point out an important fact while we operate in this way: having or not having the comparisons with the dropped alternative does not alter the other comparisons, and so we observe that *the evaluations on the alternatives are not modified if we drop, or possibly add, alternatives.* Similarly, a Condorcet winner remains the winner even if we drop alternatives or change the preferences among the other alternatives. We shall come back to this important property later.

In this case, with only three alternatives, the procedure is very simple. Only C and B remain, and, for them, we already have the direct comparison available. Hence, the collective ranking is $A \Rightarrow C \Rightarrow B$.

4.3 Condorcet Cycles

Of course, things are not so simple as they might appear at first sight. Indeed, there may be no Condorcet winner, in other words, it may happen that each alternative is defeated by at least one other alternative. It seems reasonable that this may occur, but the baffling aspect is the following: take any alternative and consider the alternative by which it is defeated (this must exist if there is no Condorcet winner). In turn, this alternative is defeated by another one, which, in turn, is also defeated by another one, and so on. Since the number of alternatives is finite, there must be a case in which, at a certain point, we find an alternative that we have already considered, and we have therefore found a preference cycle, which is indeed called a *Condorcet cycle*.

This paradoxical aspect, which Condorcet himself discovered, shows that, while preference cycles are not admitted individually, as we have already discussed, contrastingly, they may occur collectively, due to the different majorities that occur from comparison to comparison. The paradigmatic case, with three electors and three alternatives, is as follows:

$$A \to B \to C, \quad B \to C \to A, \quad C \to A \to B,$$

for which there is the cycle $A \Rightarrow B \Rightarrow C \Rightarrow A$. Indeed, A defeats B two times out of three, B defeats C two times out of three and, finally, C defeats A two times out of three. This is a case of total symmetry among the alternatives, and there can be no method that expresses a winner with the available information.

With a large number of electors, the probability of symmetries among the alternatives is very low. Hence, we may wonder if, barring symmetries, it is possible to use the available information to establish a winner anyway, even in the presence of cycles. Imagine the following example, which has been obtained by slightly modifying the previous example:

$$
\begin{aligned}
&\text{2 electors:} \quad A \to B \to C \\
&\text{2 electors:} \quad C \to A \to B \\
&\text{1 electors: :} \quad B \to C \to A
\end{aligned}
$$

This yields the Condorcet table

	A	B	C
A		**4**	2
B	1		3
C	3	2	

There is no Condorcet winner, because each alternative is defeated by some other alternative and the cycle $A \Rightarrow B \Rightarrow C \Rightarrow A$ exists.

Over time, several methods have been proposed for overcoming the deadlock caused by the presence of a cycle [54, 97, 104]. Lull himself devised one way to solve this problem. He counted the number of victories earned by an alternative in each pairwise comparison (so, in our tables, the number of entries in boldface is counted). The alternative with the most victories is the winner. Unaware of Lull's results, Copeland has proposed the same idea more recently [32], and so this way of solving the problem is known as Copeland's method. A Condorcet winner is necessarily a Copeland (or Lull) winner. However, this method does not always produce a winner, as our little example, in which all alternatives have one victory each, demonstrates.

The method that seems most valid is quite recent and was proposed by Schulze [85, 86]. The basic idea of the method is based on the observation that, if a cycle is present, beside the direct preference $A \Rightarrow B$ (as in the example), there is another indirect preference consisting of a sequence of preferences $B \Rightarrow C \Rightarrow A$.

How to evaluate these indirect preferences? For instance, if we evaluate the sequence $B \Rightarrow C \Rightarrow A$, we see that three electors prefer B to C and three electors prefer C to A. Thus, we can say that the indirect preference of B over A through the sequence $B \Rightarrow C \Rightarrow A$ is supported by at least three electors, and so the 'strength' of the preference is evaluated as 3. It is useful to note that we do not take into account the common electors shared by the two groups of electors (in this example, there is just one elector in common). The groups may consist of different electors. The only thing that matters is their number. Hence, given a sequence, its strength is computed as the smallest number in the various numbers of the pairwise preferences.

If there are several sequences linking two alternatives, it seems reasonable to evaluate the strength of a preference (either direct or indirect) as the largest value among all possible sequences. This computation, which,, at first sight seems quite complex, can be done very efficiently on a computer. Let us consider the following example, slightly more complex than the previous one (nine electors and four alternatives):

$$3 \text{ electors:} \quad A \rightarrow B \rightarrow C \rightarrow D$$
$$3 \text{ electors:} \quad D \rightarrow C \rightarrow A \rightarrow B$$
$$2 \text{ electors:} \quad B \rightarrow C \rightarrow A \rightarrow D$$
$$1 \text{ elector:} \quad D \rightarrow A \rightarrow B \rightarrow C$$

From these data, we get the following Condorcet table:

	A	B	C	D
A		7	4	5
B	2		6	5
C	5	3		5
D	4	4	4	

which shows the cycle $A \Rightarrow B \Rightarrow C \Rightarrow A$. There are four electors that directly prefer A to C, but the strength of the indirect sequence $A \Rightarrow B \Rightarrow C$ (in this example, there are no other sequences, but, in general, there could be many more) is the minimum between 7 (preference $A \Rightarrow B$) and 6 (preference $B \Rightarrow C$), and so it is 6. This value is higher than the five electors of the direct preference $C \Rightarrow A$, and thus the 'strength' of A versus C is higher than that of C versus A. By completing this type of computation for all pairs of alternatives, we get the following table, which we may call a *Schulze table*, in which the numbers represent the strength of the alternative of the row with respect to the alternative on the column (for each pair, we denote the winning alternative, i.e., the one with more strength, in boldface)

	A	B	C	D
A		7	6	5
B	5		6	5
C	5	5		5
D	4	4	4	

Thus, the choice falls on A. To complete the ranking, we easily see that, in this example, $A \Rightarrow B \Rightarrow C \Rightarrow D$. It is not difficult to see that a Condorcet winner is also a winner for the Schulze method.

We have to add that the Schulze method might also not produce a winner, because there may be ties, even if, as already stated, there is a very low probability of this with many electors. There are various proposals for a way out from the impasse [85, 86].

4.4   Principle of Independence of Irrelevant Alternatives

So far, it seems that we have solved a deadlock situation in a satisfactory way. However, we have also lost an important property. Let us examine a situation in which, for some reason, an alternative is no longer available. For instance, in a musical contest, a candidate might feel sick and not be able to play. The deletion of an alternative from the Condorcet table can be done easily by simply dropping the row and the column corresponding to that alternative. We don't have to worry about the other numbers in the table, because they have been obtained independently of the presence of the deleted alternative.

If B withdraws in the previous example, a surprising thing happens: C becomes the Condorcet winner. We recall that A was not the Condorcet winner, but turned out to be the winner only after the Schulze method was applied. A Condorcet winner remains the winner independently of the cancellation of alternatives, but, in our case, there was no Condorcet winner with four alternatives.

What about a withdrawal of B after having applied the Schulze method and proclaimed A the winner? Given the ranking $A \Rightarrow B \Rightarrow C \Rightarrow D$ that we have computed, we would very likely continue to consider A the winner, thus violating the Condorcet criterion that we had decided to consider compelling.

How is this paradoxical situation possible? While a Condorcet table does not depend on other choices, a Schulze table does depend on the other comparisons. A prevails over C thanks to the presence of B. It is the seven preferences of A over B and the six preferences of B over C that give us the strength six for A versus C. But, if B disappears, these numbers also disappear and A is defeated against C.

Clearly, this is not a minor issue. Besides, different outcomes may arise from situations other than missing alternatives. If, for instance, the two electors that have expressed the ranking $B \to C \to A \to D$ do not change their opinion between A and C, but do change their opinion between B and C, thus producing the new ranking $C \to B \to A \to D$, the Condorcet table is modified as

	A	B	C	D
A		7	4	5
B	2		4	5
C	5	5		5
D	4	4	4	

from which we see that C is the Condorcet winner and the new collective ranking is $C \Rightarrow A \Rightarrow B \Rightarrow D$ (both for Condorcet and for Schulze). Hence, two electors have changed their preferences for the alternatives B and C, but nobody has changed their preference between A and C, and yet the collective evaluation between A and C has been modified.

The fact that the collective preference between two alternatives may change only because another alternative is or is not present, or some individual preferences of other alternatives are changed, is so important as to deserve formal expression as a new criterion to be observed (if possible):

Criterion of the Independence of irrelevant alternatives: *The collective preference within any subset of alternatives must be invariant with respect to any change of individual preferences outside of the subset.*

Therefore, Condorcet's method, when it produces a winner, does respect the Criterion of the Independence of irrelevant alternatives, while the Schulze method, which yields a winner when Condorcet is not able to do so, does not respect the criterion in general.

4.5   Choice Among Many Motions

Condorcet's method does indeed have the purpose of choosing a motion among many motions that have been proposed in the discussion of an assembly. It is worthwhile once more to insist upon this issue, because it frequently happens in politics that many proposals are presented, but none reach the necessary majority for approval. The recent events in the British parliament that was faced with many proposals concerning the Brexit affair are a clear example of how politics can go into a deadlock with the usual electoral method.

Anyone who is familiar with sport might think of proposing a method that is widely employed in sport competitions, that is, a direct elimination championship, with 'matches' between the various alternatives, quarter finals, semifinals and a final. Unfortunately, it is not a productive idea.

Condorcet's method is already based on direct 'matches', but between all pairs of alternatives, not just some alternatives, as happens in direct elimination championships. If there are no Condorcet cycles, the alternative that wins in the final is also the Condorcet winner. However, the possible presence of cycles, hidden by the fact that not all matches are played, makes the final outcome very dubious.

Still taking our cue from sport competitions, we might think of a round-robin tournament, like in many football championships, in which all possible matches are played. This is similar to Condorcet's method, which compares each alternative to all others. The football championship foresees a score for each match played. If we forget about the possibility of draws, we see that the final score corresponds to counting the number of victories (multiplied by three), and the winning team is the one with the most victories. But this is just Lull's method, later reinvented by Copeland to solve a cycle!

Hence, for the reasons that we have already explained, adopting the scheme of a football championship would produce some drawbacks. Of course, the question of choosing one motion among many others is by no means simple. We shall have the opportunity to get into a deep discussion in the following chapters.

When an assembly is faced with many motions, the question should usually be foreseen in the assembly rules, even if people try to avoid it by reducing the motions to just two. If there are two motions, the question is simple (even if the motion is unique, there are actually two motions, either to accept the motion or to reject it): one motion is voted against the other, the votes in support of one motion and those in support of the other are counted and the motion with the most votes is approved (some vote mechanisms are conservative and, when the motion is unique, abstentions are counted among the 'no' votes).

Reducing the motions to only two may produce 'interesting' results. Let us assume that there are three motions A, B and C. Two motions are chosen, for instance, A and B, and are voted upon. If A passes, then A is next voted against C. If A passes again, the question is decided in a clear way: A is preferred to both B and C, and is therefore the Condorcet winner. However, if C passes in the second vote, the question is often solved in an assembly by invoking a transitivity principle: if C is better than A and A is better than B, then C must be better than B, and so C is the motion to choose.

Invoking a transitivity principle is, of course, a mistake, because we have seen that cycles may exist, and so, even if $A \Rightarrow B$ and $C \Rightarrow A$, this by no means implies $C \Rightarrow B$; on the contrary, it might be $B \Rightarrow C$. Hence, a third vote should be carried out between B and C. The chairperson of an assembly who is supposedly in favor of C, but has the suspicion that C may be defeated by B, might first put to vote first A versus B so as to get rid of B, and then A versus C. If the assembly is unaware of Condorcet cycles (quite likely), the chairperson could slyly invoke the transitivity principle and close with the approval of C.

Chapter 5
Borda

In 1770, the French officer Jean-Charles de Borda[1] proposed a method for electing
the members of the French Academy of Sciences [23]. His method would produce
a ranking among the candidates and would also clearly determine a winner, barring
possible ties. Borda's method is still very popular and currently remains in widespread
use. His method is simple, but, as frequently happens with things that are too simple,
it also presents many negative aspects.

It is for this reason that we wrote in the previous chapter, that the modern science
of Social Choice began with the essay by Condorcet, even if it appeared a few years
later.

As Condorcet's method had a precursor in Lull's method, Borda's method also
had a precursor. In 1433, Nicholas aus Kues (Nicolaus Cusanus) proposed a method
for electing the emperor of the Holy Roman Empire that is almost exactly Borda's
method [49]. Cusanus was aware of Lull's writings (in fact, it is thanks to his copies
that Lull's texts were discovered) from which he has taken some ideas. Also Cusanus'
work was soon forgotten, and it is certain the Borda was not aware of it.

5.1 Rankings and Points

In Borda's method, as in Condorcet's method, each elector gives a complete ranking
among all alternatives. The two methods differ in the use of the rankings. In Borda's
method, each individual ranking has the purpose of assigning points to the alternatives. If there are n alternatives, in each ranking, $n - 1$ points are assigned to the
first ranked alternative, $n - 2$ to the second one, and so on, up to the last alternative,

---

[1] Jean-Charles, chevalier de Borda (1733–1799), showed a precocious interest in science, which he
then studied as an officer, first in the Army and then in the Navy. He made important contributions
to many branches of Physics.

© Springer Nature Switzerland AG 2020
P. Serafini, *Mathematics to the Rescue of Democracy*,
https://doi.org/10.1007/978-3-030-38368-8_5

which is given zero points. Then, all points are summed for all rankings and the collective ranking is formed according to the final sum.[2]

Let us reconsider the example on Sect. 4.2 of the previous chapter with five electors and three alternatives:

$$1 \text{ elector:} \quad A \to B \to C$$
$$1 \text{ elector:} \quad A \to C \to B$$
$$2 \text{ electors:} \quad C \to A \to B$$
$$1 \text{ elector:} \quad B \to A \to C$$

With three alternatives, the points to be assigned are 2, 1 and 0. So, the assigned points are (one column for each elector and the sum in the last column)

	1	2	3	4	5	Totals
A	2	2	1	1	1	7
B	1	0	0	0	2	3
C	0	1	2	2	0	5

The collective ranking is therefore $A \Rightarrow C \Rightarrow B$, the same as we had obtained with Condorcet's method. Just note that we can also obtain the final score directly from the Condorcet table by simply summing the numbers in each row (the proof is not difficult). This was the Condorcet table for the example.

	A	B	C
A		4	3
B	1		2
C	2	3	

In Borda's method, the points to be assigned, namely, all integers from $n - 1$ to zero, are fixed a priori and are assigned only by looking at the ranking of the alternatives. In this regard, even if Borda's method makes use of numbers differently than does Condorcet's, it is nonetheless the order that matters in both cases. In other words, for each elector, the points do not reflect the higher or lower importance of an alternative, but only its position in the ranking.

We might think that we are loosing important information if we do not also introduce a quantitative aspect into the numerical ranking. Actually, the use of numerical quantitative information presents serious problems that cannot be easily overcome.

---

[2]Alternatively, one could assign the same points as the ranking places, one point to the first, two to the second, and so on. The collective ranking is formed by looking at the smallest numbers. Of course, this is entirely equivalent to Borda's method. This score was used in the Tours de France from 1905 until 1912.

Suppose that we modify the method by giving the electors the possibility of using any number, say, from 0 to 10, in order to evaluate each alternative. This way of voting is called *Range voting*, and it enjoys some popularity, mainly within social networks.

By working in this way, we take for granted that a certain number has the same meaning for each elector. But this is almost never true, and so distortions arise in the collective evaluation. Even when the number of electors is small, for instance, in juries for musical contests or in committees that evaluate projects, it never happens that judges meet together to discuss the exact meaning of the evaluation numbers.

Another aspect is related to the fact that, for the simple reason that electors are free to use any number, they could use this freedom not only to give the highest score to their preferred alternatives, but also lower scores to certain other alternatives with the goal of preventing their victory. In this case, we speak of *strategic voting*. Strategic voting means that we vote differently from our true preferences in order to achieve our goals. Also, strategic voting causes distortions in the collective evaluation.

The last aspect, perhaps the most important and generally overlooked, concerns the use that we make of the numbers when we measure something. It is worth devoting some words to this issue. We quote from [56]: "When measuring some attribute of a class of objects or events, we associate numbers with the objects in such a way that the properties of the attribute are faithfully represented as numerical properties". This means that, after having transformed objects into numbers, the arithmetical operations that we will inevitably perform on these numbers must make sense with respect to the objects that we have measured.

When we take the average of a set of numbers, we should be aware that we take for granted that the difference between, say, a and $a + 1$, has the same value as the difference between b and $b + 1$, for any b. Otherwise, the increment between two values cannot be compensated by the same decrement between two other values. In other words, the scale must be uniform to get meaningful averages.

For instance, scores used in exams rarely exhibit a uniform scale (and are very likely different from teacher to teacher). Yet, we continuously compute averages of exam scores, as if it were the most natural and correct thing in the world. Refer to the interesting observations in [15]. While it makes sense to compute the average height of a group of people, because 'the properties of height are faithfully represented as numerical properties', it would make no sense, for instance, to compute their average IQ, because the IQ scale is a conventional non-uniform scale. In general, when a scale is based on subjective evaluations, averages should not be taken into consideration.

5.2 Difficulties in Borda's Method

Therefore, Borda's method does not make the mistake of assigning quantitative values to the numbers. However, other problems arise. Let us consider the following situation with five electors and three alternatives:

$$2 \text{ electors:} \quad A \to C \to B$$
$$2 \text{ electors:} \quad C \to B \to A$$
$$1 \text{ elector:} \quad A \to C \to B$$

that generates the Condorcet table

	A	B	C
A		3	3
B	2		0
C	2	5	

from which we get the Borda score by summing the points on each row: 6 points for A, 2 for B and 7 for C. Hence, according to Borda, C is the alternative to choose. However, it is A that satisfies the Majority criterion (three electors out of five prefer A) and also the Condorcet's criterion. So, Borda's and Condorcet's methods disagree.

But there is one more thing. If we drop the alternative B in the example, we get 3 points for A and 2 points for C, and so A is the winner, whereas, with the presence of B, the winner was C. This means that the criterion of Independence of irrelevant alternatives is violated.

In conclusion, if we hold these criteria to be true, we have to admit that Borda's method, which may violate the Majority criterion, Condorcet'a criterion and also the criterion of Independence of irrelevant alternatives, is very weak. The reader can also examine the outcome produced by Borda's method for the examples on Sects. 4.3 and 4.4. In both cases, the Borda winner is A.

In spite of these defects, Borda's method is widely used, even if with some variations. In Formula One, points are assigned to the first ten drivers in each race. The points are not from 10 to 1, as it would be with Borda's method, but are rather 25, 18, 15, 12, 10, 8, 6, 4, 2, 1. One additional point is assigned to the driver that does the fastest lap. As can be seen, the points are not uniformly distributed. The first positions are rewarded. Therefore, a quantitative element is introduced that is not present in Borda's method. This variant also presents all of the drawbacks of Borda's method and thus it is no wonder that the scoring method has been changed many times over the years.

To understand the degree of inconsistency that exists in carrying out arithmetic operations on the numbers obtained from a ranking, one can evaluate the score obtained by Lewis Hamilton in the Formula One 2017 (in 2017, the extra point for the fastest lap has not yet been introduced). The final score was 363 points in 20 races. By taking the average, one obtains 18.15 points for each race, which amounts to saying, looking at the scoring method, that the pilot was, on average, slightly better than second. If we look at the single races in detail, we see that Hamilton won 9 races, came in second four times, fourth four times and fifth, seventh and ninth only once each. What we can say is that, in 9 races, he was better than second, and in 7 races, he was worse than second. So, was he eventually better or worse than second? We can't say anything, because we need to quantify the case 'worse than second' for each of the 7 races and, in any case, this quantification would be arbitrary. If, for instance, we use Borda points (10 points to the first, 9 to the second, and so on), we obtain 166 points, on average, 8.3 points per race, that is to say, worse than second.

In any case, the outcome of the Formula One 2017 was not problematic. Hamilton also turned out to be a Condorcet winner (11 times better than Vettel, 12 better than Bottas and 15 better than Räikkönen and, of course, better than the others by larger margin), but not a majority winner, because he won less than half of the races.

It is useful to devote some more words to these methods to determine the collective winner, making explicit reference to Formula One. Suppose that a driver A wins 11 races out of 20, that is, more than half, but, in all other 9 races, he pulls out, or he gets zero points. According to the Majority criterion (and so also according to Condorcet's criterion) he should be the world champion, independently of the scores of the other drivers. However, let us further suppose that there is a driver B who has been skillful enough to come in second in those races won by A and first in the other 9 races. Who should be the world champion? According to the actual scoring method (neglecting the extra point for the fastest lap), A would get 275 points and B 423, and so B would be the champion. Even without invoking any special scoring method, it seems that B deserves the victory, because, on average, he has performed much better than A.

So, what is the right thing to do? Adopting the Majority criterion or a method like Borda's? If we reflect, we have to note that the word 'right' is meaningless unless we specify a priori a criterion that establishes what is right and what is wrong.[3] If we have decided a priori that the Majority criterion must be satisfied, then we are not allowed to complain if driver A becomes world champion, because, according to the Majority criterion, it is 'right' that A wins.

This example drawn from a sporting context may be disconcerting. It has to be said that it is an extreme example whose probability of occurring in reality is practically zero. Moreover, it has to be added that, given any method of voting, we can build up tailor-made examples that will make that method very implausible.

However, let us reason a little bit further on this example. We can transfer it into the political realm, and maybe we will be able to accept its outcome with less difficulty. The 11 races won by A become 55% of electors that prefer A over all others. The other 9 races won by B become 45% of electors that prefer B and also abhor A (the pilot A got zero points in those races). With the present voting mechanism, in which we assign only one vote and not the entire ranking, we only see that A gets 55% of votes and B 45%. The fact that A is abhorred by 45% of the electors is invisible, and thus the choice falls on A. This outcome does not seem strange. Clearly, this also depends on the significantly reduced amount of information that is used in Single vote systems.

What about using more information, for instance, taking into account the fact that A is despised by a large part of the electorate, while B is liked, more or less, by almost everyone? Thus, by adopting the Majority criterion, A would be the winner in all cases, but by adopting a Borda-like criterion, B would be the winner.

---

[3]"[...] for there is nothing either good or bad, but thinking makes it so." *Hamlet*, Act II, Scene II.

This seems like a dilemma without any way out, because robust criteria like Majority and Condorcet's criteria produce a result that does not seem to reflect the will of the society, whereas a criterion like Borda's, albeit a flawed one, produces, at least in this case, a satisfactory result. We shall see, in Chap. 8, a (partial) way out of this dilemma.

5.3   Difficulties in Range Voting

We have already touched upon the variant of Borda's method, called Range voting, in which each elector assigns a point to each alternative within a preset range (for instance, from 0 to 10), and then one sums up all points obtained by each alternative (or one can take the average, which is equivalent if each alternative is scored by the same number of electors). We have already stressed that this method presents the serious drawback that the same number does not have the same meaning for each elector, and so we sum non-homogeneous values. Furthermore, if the scale is not uniform, carrying out averages makes no sense.

We also mentioned that another non-negligible defect of Range voting consists in its manipulability, since electors can assign points that are quite different from the ones that they have in mind, for the sake of making some particular candidates win or lose. For instance, three electors (1, 2 and 3) might have in mind the following true scores in regard to three alternatives A, B and C:

	1	2	3
A	8	7	8
B	10	10	7
C	7	8	6

that make B the winner with 27 points. However, the third elector, who prefers A to the others, even if only by a small margin, assigns the maximum score to A and 0 to B and C to avoid the risk of B or C winning. The scores are therefore modified as

	1	2	3
A	8	7	10
B	10	10	0
C	7	8	0

making A the winner. However, if electors 1 and 2 suspect that elector 3 is assigning scores to make B lose, they may behave in the same way, assigning the scores

	1	2	3
A	0	0	10
B	10	10	0
C	0	0	0

At this point, we have come back to Single vote, and the possibility of exploiting more information to better evaluate the alternatives is frustrated by strategic voting.

Beside these defects, even if we assume that electors have the same meaning in mind for the numbers within the range and they do not vote strategically, it is easy to see that the method does not satisfy the Majority criterion, and therefore nor does it satisfy Condorcet's criterion. The criterion of Independence of irrelevant alternatives is, contrastingly, satisfied. If we drop some alternative or simply modify some of its points, the score of the other alternatives is clearly unchanged. However, even if this criterion is satisfied, the other defects advise against using Range voting.

Chapter 6
Simple Majority and Run-Off Voting

With the present voting system, in which each elector gives a vote to their most preferred alternative, it frequently happens that no alternative reaches the absolute majority. So, it would seem natural to choose the alternative that is preferred by most electors, the so-called *simple* or *relative majority*. Let us call this method of choosing *Simple majority criterion*. In regard to this issue it is paramount to be crystal clear: *by operating according to the Simple majority criterion, we make a fundamental mistake.*

6.1 Problems with Simple Majority

In order to understand what constitutes the mistake, we have to bear in mind that, even if electors express only one vote, regardless, in their minds, a complete judgement over all alternatives does exist, and, in particular, a ranking as well. Hence, we must evaluate what difference there can be between the outcome brought about by using a single vote with the Simple majority criterion and the outcome brought about by using the whole ranking. Let us consider the following situation with three alternatives and seven electors:

$$1 \text{ elector: } A \to B \to C$$
$$2 \text{ electors: } A \to C \to B$$
$$2 \text{ electors: } C \to B \to A$$
$$2 \text{ electors: } B \to C \to A$$

A is the first choice of three electors out of seven, which corresponds to almost 43%, while both B and C are preferred by two electors, that is, 28.5%. With Single vote, these percentages would be the vote outcome and, if we choose according to Simple majority (or we assign a majority bonus to the coalition that reaches 40%, but not 50%, as it has been done in Italy), the choice would fall on A.

Now, we analyze the vote in depth by explicitly taking into account the rankings of the electors. The Condorcet table is

© Springer Nature Switzerland AG 2020
P. Serafini, *Mathematics to the Rescue of Democracy*,
https://doi.org/10.1007/978-3-030-38368-8_6

	A	B	C
A		3	3
B	4		3
C	4	4	

yielding the Condorcet winner C, followed by B, whereas A is last. The Borda score gives the same result: 8 points for C, 7 for B and 6 for A. So, for both Condorcet and Borda, we have the collective ranking $C \Rightarrow B \Rightarrow A$.

Choosing according to the Simple majority criterion may therefore result in the choice of the alternative that would be last if we used more information!

What has been highlighted in the example does not correspond at all to an anomalous situation. The alternative A could correspond to a party that is quite different from B and C, whereas B and C could be similar parties. Together, they share 57% of votes, but, since they stand for election separately, they get fewer votes than A, and thus A is chosen, leading to an outcome that does not reflect the unexpressed will of the electors.

One might argue that parties B and C should have formed a coalition to stand for election and, in some sense, the A's victory is their 'fault'. However, why should we deprive the electors of the possibility of distinguishing between the two parties without penalizing both the parties and the electors simply for this reason?

The Simple majority criterion is currently employed in many countries. The uninominal system in which each district sends the most voted-for candidate to the parliament is an example of the Simple majority criterion. If there are only two alternatives, then the simple majority is also the absolute majority, but with more than two alternatives, we face the problems that we have previously seen. The uninominal system employed in nations with a history of a wide spectrum of political choices may give rise to very distorted outcomes.

The Presidents of many Regions in Italy are elected according to the Simple majority criterion. Italian mayors of municipalities with fewer than 15,000 people are elected by Simple majority. It is argued that these are little communities, and therefore it does not make sense to trouble the electors with a possible second ballot, as would be prescribed for municipalities with more than 15,000 people. However, the practical effect is that, to avoid vote-splitting between 'similar' candidates, the candidates for the mayoral position are decided a priori by the parties, and so the electors are faced with prearranged choices. But it is not only in smaller territorial units that Simple majority is employed. The uninominal seats of the last Italian political elections (2018) had a number of people per seat ranging from 200,000 to 300,000 people (see Sect. 13.3 and Fig. 13.4).

Besides the theory, there are also historical precedents that show how dangerous it can be to adopt the Simple majority criterion. In July 1932, the German political elections gave the Nazi Party the simple majority, 230 seats out of 608, 37.82%, but no government could be formed. The elections were thus repeated in November and the Nazi Party, though losing seats, again obtained the simple majority with 196 seats out of 584, 33.56%. But this time, the government was formed and, on January 30, 1933, Hitler was appointed Chancellor. As a result of the fire at the Reichstag and the subsequent emergency decree that suppressed many rights, it was possible for Hitler to call for new elections in March 1933, the result of which was again the simple majority for the nazis, 44% of votes with 288 seats out of 647, 44.51%. But this time, the margin for the absolute majority had been reduced, and an alliance with an extreme right party was sufficient to get it. With the banishment of the Communist Party the majority became larger, and so it was possible for Hitler to promulgate an ordinance that, on March 24, 1933, gave him full powers. The next step was the banishment of all parties (except the Nazi Party) on July 14, 1933. The rest of the story is well known.

6.2 Run-Off Voting

With the simple mechanism of giving one single vote, there still exists a possibility of making a choice without necessarily resorting to Simple majority. It is just a matter of voting again with a smaller number of available alternatives. This second ballot, called *Run-off voting*, is not a repetition of the first vote, but a new vote, and, as a matter of fact, its purpose is to better probe the opinions of the electors, by further revealing, though always partially, the ranking that electors have in their minds.

In the second ballot, all alternatives but the first two are excluded. In theory, one should exclude all alternatives that, even if united together in a coalition, would not be able to get the absolute majority. To this purpose, one could scan the list of alternatives, starting from the least voted-for and stopping when, by adding one more alternative, the total of votes would pass 50%. These alternatives should be excluded from the second ballot. However, this procedure would imply that there could be more than two alternatives admitted to the second ballot, and this, in turn, could require a third ballot. Actually, there are always two and only two alternatives admitted to the second ballot. Almost always, the two most voted-for alternatives together exceed 50% and even if this is not so, the choice is limited to two alternatives, both for cost reasons and also because it is not convenient to obligate the electorate to fill out more than two ballots, otherwise people lose interest in voting, with a consequent vote distortion from one ballot to the next.[1]

So, the Run-off voting method excludes some alternatives. It is therefore important to see whether the criterion of Independence of irrelevant alternatives is violated by the method of election that we want to adopt. If we have in mind Condorcet's method which does not violate the criterion when there are no cycles, then we may count on a robust result. However, Run-off voting only partially reveals the electorate's preferences, so that the winner in the second ballot might be not the Condorcet winner.

---

[1] Anyone who finds delight in math may grapple with showing that, by having many ballots available, the number of necessary ballots, according to the explained rule, cannot exceed the value $\lceil \log_2 n \rceil$.

Let us again consider the example on Sect. 6.1. The alternatives B and C are second with equal merit, and so we are not able to decide who should be excluded. With many electors, the probability of equal votes is almost negligible, and so we don't have to worry about this possibility. Thus, let us modify the example by increasing the number of electors and slightly modifying the proportions of the data:

$$
\begin{aligned}
&100 \text{ electors: } A \to B \to C \\
&200 \text{ electors: } A \to C \to B \\
&199 \text{ electors: } C \to B \to A \\
&201 \text{ electors: } B \to C \to A
\end{aligned}
$$

yielding the following Condorcet table:

	A	B	C
A		300	300
B	**400**		301
C	**400**	**399**	

We get the same result as before. C is both the Condorcet winner and the Borda winner. However, in term of single votes, we have: 300 votes (42.86%) for A, 201 (28.71%) for B and 199 (28.43%) for C. So, A and B go to the second ballot, whereas C, the Condorcet winner, is excluded!

Hence, in Run-off voting as well, we get an 'unfair' result (if we adopt Condorcet's criterion as our fairness criterion). The reason lies in the fact that we only have partial information. In any case, by having the possibility of a second ballot, we end up with a better result than the one obtained by adopting the Simple majority criterion. By excluding C, 400 electors prefer B to A and only 300 prefer A to B, so B would be chosen. It is not possible to choose C, which is the choice that is consistent with the will of the electorate. However, A, the last in the collective ranking, is not chosen.

6.3 Instant Run-Off Voting

It is not really necessary to materially repeat the election to apply the idea of Run-off voting. We may carry out a virtual second ballot by asking the electors to indicate not only their first choice, but also their second choice, their third choice, and so on. This method of voting is called *Instant run-off voting*, as well as *Single transferable vote*, *Ranked choice*, or *Preferential vote*. This way, the second choice is expressed immediately and not just in the second ballot, when the political scenario might have changed. Furthermore, while, in the Run-off voting, all alternatives are excluded but the first two, in Instant run-off voting, the alternatives are eliminated one at a time.

The way it works is simple. At first, only the first choices of the electors are taken into account. If no alternative gets the absolute majority, then we must exclude some alternatives. The method that is usually employed excludes one alternative at a time, clearly the least voted-for. The votes are counted again, and wherever the first choice was the excluded alternative, it is the second choice that is counted. The procedure goes on until some alternative gets the absolute majority.

It is clear that each elector should indicate at least as many choices as the number of possible eliminations from the list. The procedure tends to be demanding for the electors in the presence of many alternatives. However, the amount of information that an elector is supposed to give is not larger than that required for Condorcet's method.

As we have just seen from the previous example, even if the type of information is similar to what is used in Condorcet's method, the method is different and the outcomes can be different. A variant of the method proposed by Coombs [31] requires giving the full ranking of alternatives and, in turn, excludes not the least voted-for first choice, but the most voted-for last choice.

In the previous example, A, B and C are the last choice for 400, 200 and 100 electors respectively. So, Coombs' method excludes A. Between B and C, C wins. In this case, the winner is also the Condorcet winner, but it is not difficult to show other cases in which this circumstance does not necessarily happen, like in the following example:

$$3 \text{ electors: } A \rightarrow B \rightarrow C$$
$$3 \text{ electors: } A \rightarrow C \rightarrow B$$
$$1 \text{ elector: } \ B \rightarrow A \rightarrow C$$
$$3 \text{ electors: } B \rightarrow C \rightarrow A$$
$$1 \text{ elector: } \ C \rightarrow A \rightarrow B$$
$$2 \text{ electors: } C \rightarrow B \rightarrow A$$

resulting in the following Condorcet table:

	A	B	C
A		7	7
B	6		7
C	6	6	

There is no majority winner. A is both the the Condorcet and Borda winner, but would be excluded by Coombs' method!

Chapter 7
Impossible Wishes

In the previous chapters, we have seen several examples in which it is difficult, if not impossible, to attain a satisfactory choice. Condorcet's and Borda's methods, with their variants, do not always provide a solution to the problem. The issue is certainly very complex. The citizens want an electoral system that is able to satisfy some important questions. Unfortunately, Social Choice theory is scattered with negative results, showing that there are some desirable guarantees, which cannot, however, be fulfilled all at the same time. Lawmakers need to be aware of this inescapable fact in order to understand in advance what can or cannot be achieved from an electoral system.

7.1 Arrow's Impossibility Theorem

The American mathematician Kenneth Arrow,[1] in his doctoral thesis from 1950 [3, 4] faced the problem of defining a method that could lead, in very general terms, to a satisfactory choice. His goal was to find a Social welfare function (which can also be called a Collective choice function), that could aggregate all individual choices into a collective or social choice.

We need to define the problem in a precise way. On one side, we have to define exactly what we mean by 'choice', both individual and collective, and on the other side, we have to identify criteria for the social choice to mirror the individual choices in the best possible way. Hence, there are *individuals* (for instance, electors, jurors in a contest, Formula One races, etc.) and *alternatives* (i.e., parties, candidates, competitors, drivers, etc.). Each individual is able to define in regard to each pair of alternatives, whether he/she prefers A to B ($A \rightarrow B$), or B to A ($B \rightarrow A$), or is

---

[1] Kenneth Joseph 'Ken' Arrow (1921–2017) was born in New York to Jewish Romanian parents. The Great Depression pushed the family into poverty, but, in spite of this, Arrow was able to successfully complete his studies. In 1972, he was awarded the Nobel Prize in Economics for his Impossibility Theorem.

© Springer Nature Switzerland AG 2020
P. Serafini, *Mathematics to the Rescue of Democracy*,
https://doi.org/10.1007/978-3-030-38368-8_7

Fig. 7.1 Functioning of a Social welfare function

indifferent between A and B ($A \sim B$). The only thing that we require from an individual is to formulate rational preferences, exactly as we already explained for Condorcet's method on Sect. 4.2.

Then, each individual ranks all alternatives with possible draws. For instance, with five alternatives A, B, C, D and E, an individual could express the ranking $D \to (A \sim B) \to (C \sim E)$, which amounts to saying that the most preferred choice is D, and then, secondarily, A and B, between which he/she is indifferent, with the least preferred alternatives being C and E, between which he/she is also indifferent.

Another individual could give the ranking $B \to (A \sim C \sim D \sim E)$, i.e., the most preferred choice is B and he/she is indifferent with respect to the other four choices. This ranking corresponds to the usual Single vote system: B is voted for and there is no information on the other alternatives. Hence, even if the individual has more articulate preferences in mind, like $B \to A \to C \to D \to E$, what can be seen from the vote is only $B \to (A \sim C \sim D \sim E)$.

So, we can look at the usual electoral method in two alternative ways: 1) as a complete ranking that, however, has the sole purpose of assigning a vote to the top alternative in the ranking; 2) as a preference of the kind $B \to (A \sim C \sim D \sim E)$ as a particular case of Arrow's scheme.

The set of all rankings expressed by the electors in the society is called the *social profile*. We also want the collective choice, derived from the social profile, to be a ranking among the alternatives with possible ties.

What we want to find is a Social welfare function, that is, a choice function (we can call it a method, a mechanism, an algorithm, as we like) that takes as its input any social profile and gives as its output the corresponding collective ranking (see a schematic representation in Fig. 7.1). Of course, the function cannot be arbitrary, but must be built up to correspond as faithfully as possible to the individual rankings. For this purpose, Arrow identifies four criteria, so 'obvious' that any choice method should abide by them. We report the criteria that Arrow presented later [5], which are more general than those presented in [3, 4].

(1) **Universality of the domain**. No restrictions are allowed on the social profile. In other words, the choice function must produce an output no matter how different the individual rankings are. The criterion sounds somewhat obvious. However, it is necessary to state it explicitly, because a choice function must work when it is applied not only to ideologically compact societies, but also to very articulate ones. If all people had the same opinion, the problem of finding a social choice function would be trivial. The criterion is also violated when some alternatives are not admitted by

authority. For instance, this may happen when a party is declared illegal and not permitted to be on the ballot.

(2) **Unanimity**. If all individuals express the same preference between two alternatives A and B, then the choice function must also yield the same preference. This requirement is meant in an extended sense: if, for all individuals, we have either $A \to B$ or $A \sim B$, then $B \Rightarrow A$ must be excluded. It is quite natural that this criterion be respected. A choice that would produce $B \Rightarrow A$ when all individual express $A \to B$ would be very strange and unacceptable.

(3) **Independence of irrelevant alternatives (IIA)**. If only some alternatives are taken into consideration and no individual changes their preferences within these alternatives, then, no matter how the other alternatives are changed, the choice function must give the same output for the considered alternatives. For instance, if there are four alternatives and four individuals and we consider the two different social profiles

$$\begin{pmatrix} A \to B \to C \sim D \\ C \to D \to A \sim B \\ A \to D \to C \to B \\ B \sim C \to A \sim D \end{pmatrix}, \quad \begin{pmatrix} A \to D \to C \sim B \\ A \sim B \to C \to D \\ C \sim D \to A \to B \\ B \to A \sim C \sim D \end{pmatrix}$$

the social choice function must produce the same relationship between A and B (the type of relationship doesn't matter, what matters is that it must be the same), either if the society (four individuals in this simple case) expresses the first profile or if it expresses the second profile, because, once we have dropped the irrelevant alternatives C and D, the two profiles are identical and equal to

$$\begin{pmatrix} A \to B \\ A \sim B \\ A \to B \\ B \to A \end{pmatrix}$$

We have already encountered this criterion and we have seen that there are methods, like Borda's method, that do not respect it. Though less obvious than the Unanimity criterion, it does anyway seem appropriate to require that a choice function abide by this criterion: if some alternatives disappear or some individuals change their opinion on them, why should this influence the other alternatives?

(4) **Non-dictatorship**. If there exists an individual for whom $A \to B$ implies that, in the collective ranking, one also has $A \Rightarrow B$, no matter what the preferences of other individuals are, then such an individual is defined as a *dictator*. The criterion excludes the existence of a dictator in the society. We note that, in order to be a dictator, an individual must only impose his strict preferences; it is not required that the indifference also be imposed. Then, if $A \sim B$ for a dictator, the social ranking can be anything with respect to A and B. This criterion also sounds obvious.

Arrow's amazing result, known, in fact, as the *Impossibility theorem*, is that *no choice function exists that satisfies the four criteria, if there are at least three alternatives.*[2]

Hence, four reasonable criteria put together produce an impossibility result. This is disconcerting, but it is a proven mathematical fact that cannot be bypassed. A consequence of the theorem that may not be immediately evident is that the only choice function that respects the first three 'obvious' criteria is the dictatorship.[3] The result seems hopeless. However, its validity concerns a problem that is formalized as we have described, i.e., with individuals that express rankings and by meeting the four criteria. Expressing a ranking seems like a very general method of 'voting', much more informative than the current method of voting. We will see in Chap. 8 that a method of voting that also asks the electors to express judgements on the alternatives, and not only their relative preference, can overcome the Impossibility theorem.

7.2 Single Vote and Arrow

For now, we may ask which criteria are violated by the current voting systems. Almost always, it is IIA that is violated. All debates (at least in Italy) about the 'useful vote' concern the precise fact that this criterion is not satisfied. The sort of consequences that its violation may bring about was seen in the American presidential elections in 1992 and 2000. In 2000, Ralph Nader, in the controversial but ultimately decisive election count in Florida, received 97,488 votes. The difference between George W. Bush and Al Gore was only 537 votes out of almost three millions votes each (data that was obtained after many tallies and a ruling by the Supreme Court of Florida). Given their political positions, it is quite likely that Nader's votes would have gone to Gore, if the former had not run for office, making Gore the U.S. President. Analogously, in 1992, Bill Clinton won with 43% of the popular vote, whereas the remaining 57% went to George H. W. Bush and Ross Perot. If Ross Perot had not run, given his rightist position, the winner would have been Bush.

It is useful to devote some more words to the current voting system with respect to Arrow's criteria. We mentioned that Single vote can be viewed in two alternative ways, either as a complete ranking $B \rightarrow A \rightarrow C \rightarrow D \rightarrow E$ (in the elector's mind) or as a particular ranking $B \rightarrow (A \sim C \sim D \sim E)$ (that which is determined from the vote). So, what happens if we drop the alternative B?

In the first case, the alternative A becomes the first choice and the vote goes from B to A. The previous discussion assumes that, if an alternative is missing, the elector keeps on voting, but for the second alternative. In this case, the violation of IIA is manifest.

---

[2]The proof requires a solid mathematical knowledge, and it is not possible even to outline it here.

[3]Actually, in the theorem proof, this property is proven first and the theorem thesis comes from this property.

In the second case, dropping B from $B \to (A \sim C \sim D \sim E)$ means that the elector no longer has a preferred alternative, and so he/she is not casting his/her vote. If those people that voted for the irrelevant alternative no longer vote, there can be no surprises for the other alternatives, which remain with the votes that they had. So, IIA is not violated. Which criterion is thus violated? It is the first criterion, Universality of the domain, since we restrict the possible rankings to rankings of the type $B \to (A \sim C \sim D \sim E)$.

7.3 Condorcet and Arrow

In general, Condorcet's method does not comply with IIA. It does if there exists a Condorcet winner, but cycles may exist, and this is not an uncommon situation. Charles Dodgson [34] (better known as Lewis Carroll) noted that strategic voting tends to generate Condorcet cycles. Hence, it is more likely that one will find cycles instead of a Condorcet winner in a real election.

Then, by considering the alternatives on a cycle on more or less equal terms, we might think of solving the cycle by assigning to the collective choice function indifference among all alternatives of the cycle. But, by working this way, we violate IIA. Indeed, if we drop one alternative in the cycle, the cycle disappears and the preference relations for the other alternatives of the cycle change.

7.4 Approval Voting and Arrow

Examples of similar parties that stand for elections separately and that each get fewer votes than another, third party, separately, but would have gotten more votes had they run together, can be seen almost at every election. This phenomenon has been denoted in the literature as *vote splitting*. It would be appropriate for an election method to be insensitive to vote splitting.

One election method that has been designed specifically to avoid vote splitting is so-called *Approval voting* [25, 26]. This method is currently used, mainly in small communities, e.g., to elect the president of an association. Instead of indicating only one candidate, electors can give their vote to all preferred (or 'approved') candidates, in any number. Others candidates are not voted for. Then, all votes are summed up and the winner is the one that has received the most votes. This method seems more informative than the normal Single vote system. It can be part of Arrow's scheme, because it corresponds to a ranking of the type

$$(A \sim C \sim E) \to (B \sim D)$$

where, in the first group, there are the voted-for equivalent alternatives and, in the second group, the non-voted-for equivalent alternatives. However, we could also have (in the elector's mind) a ranking of the type $(A \sim C \sim E) \to B \to D$, and we may evaluate the Approval voting method with respect to Arrow's criteria in either way.

According to the Impossibility theorem, one of the four conditions must be violated. If we consider the first type of rankings, Universality of the domain is clearly violated because we have only a restricted set of rankings as our input [60]. The IIA criterion is violated if, by dropping all alternatives from the first group, the elector votes for the most preferred alternatives from the second group anyway. It is clear that, if we drop alternatives, we are only eliminating votes for these alternatives, but the other scores remain invariant, and so does the collective ranking. However, as with Single vote, we have to ask ourselves what the elector does when their preferred alternatives disappear. Is he/she going to abstain or vote for somebody else?

It may be interesting to see what outcome Approval voting produces with respect to the other methods that we have presented. Let us consider the following example, which is a slight variation on the one from Sect. 4.3. The number of electors has been set at 1000 so that the percentages can be seen at once:

$$350 \text{ electors:} \quad A \to B \to C \to D$$
$$300 \text{ electors:} \quad D \to C \to A \to B$$
$$250 \text{ electors:} \quad C \to B \to A \to D$$
$$100 \text{ electors:} \quad D \to A \to B \to C$$

There are no majority winners. According to Single vote, the simple majority is obtained by D, with 400 votes. A gets 350 votes, C 250 and B none. The Condorcet table is

	A	B	C	D
A		750	450	600
B	250		450	600
C	550	550		600
D	400	400	400	

from which we get the Condorcet ranking $C \Rightarrow A \Rightarrow B \Rightarrow D$. If we apply Borda's method, we get the ranking $A \Rightarrow C \Rightarrow B \Rightarrow D$ (respectively, with 1800, 1700, 1300, and 1200 points), which shows once more that Condorcet's and Borda's methods may produce different results. Noteworthy is the fact that D, despite having obtained the simple majority (40%) with Single vote, turns up at the bottom of the ranking with both Condorcet and Borda. This is a further example of the fallacy of the Simple majority criterion.

Let us now adopt Approval voting and assume, just for the sake of argument, that the electors indicate the first two alternatives in the respective rankings. So, A gets 450 preferences, B 600, C 550 and D 400. By this method, the ranking is $B \Rightarrow C \Rightarrow A \Rightarrow D$, a result that is different from both Condorcet and Borda (and D is still last).

7.5 Only Two Alternatives

It is important to stress that the impossibility stated by the theorem refers to the presence of at least three alternatives. With only two alternatives, the IAA criterion becomes void, because there are no irrelevant alternatives and there are only three criteria to abide by. A collective choice function that satisfies these three criteria can be easily constructed. For instance, it is enough to see whether $A \to B$ prevails with respect to $B \to A$, or vice versa. The collective function yields $A \Rightarrow B$ or $B \Rightarrow A$ in the respective cases. If, contrastingly, we get $A \to B$ and $B \to A$ for the same number of electors, the collective function yields $A \approx B$. With many electors, the probability of this last case is very low.

Apparently, this choice function corresponds to the Majority criterion (applied only to the individuals who express preference and not indifference). The idea of having only two alternatives is therefore very appealing, because it removes by the root all of the paradoxes and difficulties that any electoral method exhibits when there are at least three alternatives. However, it is not conceivable to force individuals to restrict themselves in their choices. If, in a country, there is a history of only two alternatives, this does not mean that the same pattern will continue in that country or that it can be reproduced in other countries by, perhaps, adopting electoral methods that force a choice between only two alternatives (as it has been attempted in Italy).

As far as the Majority criterion with two alternatives is concerned, we may add some important considerations. It seems natural to require a collective choice function to be indifferent, both with respect to the alternatives and with respect to the individuals. Going into more detail, if we switch two alternatives in all individual rankings, then, in the collective choice, the outcome must be the same, with the only difference being the switch between the two alternatives. So, no alternative is a priori favored. Analogously, if two electors switch their rankings, the collective choice must be invariant, which amounts to saying that all individuals are equal.

A social choice function that is indifferent with respect to the alternatives is called *neutral*, while it is called *anonymous* if it is indifferent with respect to the individuals. Moreover, a choice function that is both neutral and anonymous is called *impartial*.

A famous theorem by May [64] states the important result that *the only collective choice function with two alternatives that is impartial is the majority method*.

Having only two parties in competition should not convey the idea that the choice function is in any way easy to find. That which has been previously written refers to the case of two alternatives, and two parties does not necessarily also mean two alternatives. If the political proposals of the parties are articulate on a variety of themes, then, de facto, the alternatives increase in number and paradoxes emerge again. Let us consider the following situation that was proposed by Ostrogorski at the beginning of the previous century [69] and is, indeed, called *Ostrogorski paradox*. Two parties A and B have explained their political programs with respect to three themes X, Y and Z (like, for instance, economy, foreign policy and the environment). Five electors have their preferences about the various themes as shown in the table and vote for the party that is preferred in regard to at least two themes.

	X	Y	Z	vote
1	A	B	B	B
2	B	A	B	B
3	B	B	A	B
4	A	A	A	A
5	A	A	A	A

Taking into account the final vote and the Majority criterion, party B is chosen. However, if we look at the single themes, we see that party A is preferred to B on every theme, again based on the Majority criterion. If we were to vote by giving one point to the preferred party theme by theme, we would see A winning with 9 points against 6 for B. So, who should govern?

It is clear that we need to look at the problem from two different points of view. In the first case, we consider the electors, one at a time, and see that three electors out of five are more satisfied with the choice of B rather than A. In the other case, we look at the themes, one at a time, and, on every theme, we see that a majority is satisfied with the choice of A. The reality that we need to deal with is the fact that this majority is variable from theme to theme and that when we have to consider all themes together (something that is, after all, inevitable), we have to take account of this variability. Therefore, the right choice is B.

7.6 Non-manipulability of an Electoral System

The four criteria stated in Arrow's theorem identify four properties that an electoral system should satisfy. Unfortunately, as we have seen, these properties cannot be satisfied at the same time. We have also highlighted two other important, if not indispensable, properties, namely, anonymity (all electors are equal) and neutrality (there are no favorite parties). Other properties have been identified in the literature, but we will deal with only a few of them, the most important.

Can there be any difference in reality between what electors desire in their minds and what they vote for instead? The answer is clearly affirmative. Quite often we realize that what is desirable has little possibility of success, and so we vote for an alternative that is close to our wishes, but that has a larger probability of success. This way of voting is called *strategic voting* and is rightly considered a distortion of the vote.

So, we may say that an electoral system is *non-manipulable* or *strategy proof* if the best choice for an elector is to vote according to his/her own wishes.

Let us again try to put together some obvious criteria that we would like to see satisfied. For instance, we may consider Unanimity, Non-dictatorship and Non-manipulability for a voting system based on rankings (as in Arrow's theorem). A theorem by Gibbard and Satterthwaite states that *there does not exist a voting system based on rankings that guarantees Unanimity, Non-dictatorship and Non-manipulability* [43, 84].

For instance, Condorcet's method, which guarantees Unanimity and Non-dictator-ship is manipulable. As a simple example, let us imagine a music contest with three candidates and three jurors, whose true rankings are:

$$\text{juror n. 1:} \quad A \to B \to C \qquad \text{juror n. 2:} \quad A \to C \to B$$
$$\text{juror n. 3:} \quad B \to A \to C \qquad \text{juror n. 4:} \quad C \to B \to A$$
$$\text{juror n. 5:} \quad B \to C \to A$$

with Condorcet table:

	A	B	C
A		2	3
B	3		3
C	2	2	

Candidate B would be the winner if the votes were to be expressed as above. The ranking is $B \Rightarrow A \Rightarrow C$. However, juror n. 1's favorite candidate is A, and juror n. 1 wants A to be the winner. Even if he believes that B is superior to C, he switches them in his ranking, which becomes

$$\text{juror n. 1:} \quad A \to C \to B$$

leading to a different Condorcet table:

	A	B	C
A		2	3
B	3		2
C	2	3	

Now, B is no longer the Condorcet winner. Nor has A become the Condorcet winner, but now the symmetric situation no longer favors B. At this point, suspecting the strategical voting by juror n.1, juror n. 3, whose favorite candidate is B, also thinks of voting strategically and adopts the new ranking

$$\text{juror n. 3:} \quad B \to C \to A$$

with the result that the Condorcet table is

	A	B	C
A		2	2
B	3		2
C	3	3	

and thus the new Condorcet winner is C, who was the preferred choice of only one juror and was last in the true ranking! Eventually, neither A nor B win. The reader can apply Borda's method to the example and obtain the same result.

7.7 Participation Criterion

A property that should, of course, hold for any electoral method is the following: if A is currently the winner with some votes still to be counted and, in all of these missing votes, A is strictly preferred to B, then it should not happen that, after counting these votes, B becomes the winner, replacing A. We are not saying that, in these last votes, A is the most preferred choice, but only that A is preferred to B and that the first choice could be different from A. Of course, we do not expect that A will necessarily remain the winner. The last votes could favor some other alternative already high in the collective ranking that could eventually overtake A. Surely, we do not expect B to win in place of A!

This criterion is called *Participation criterion*, and its violation gives rise to the so-called *No show paradox* [39]. Maybe this criterion is not fundamental for political elections. Even if counting the votes takes some time, the electors are aware that one must wait for the counting to be concluded before looking at the final outcome. In contests and sport competitions, all jurors must cast a vote and do so all at the same time. Hence, the paradox has no possibility of arising.

However, it is important to understand how voting systems, seemingly so natural, can give rise to unexpected results that go against common sense. Many electoral systems violate the Participation criterion. However, Single vote and Borda's method don't. Indeed, these are methods that, at each counted vote, add something, and so they cannot overturn a situation in favor of B and against A if the additions always see A in a position of advantage with respect to B. Condorcet's method, when a Condorcet winner exists, admits the No show paradox under the condition that there are at least 4 alternatives and 25 electors [67]. Also, all methods that try to solve the Condorcet cycles admit the paradox. It would be tedious to illustrate complex examples that show the various cases. We limit ourselves to giving a simple example for a voting system that is close to everybody's experience, like Run-off voting.

Imagine the following situation with three alternatives and 13 electors (note that there is no Condorcet winner):

$$3 \text{ electors: } \quad C \rightarrow A \rightarrow B$$
$$4 \text{ electors: } \quad A \rightarrow B \rightarrow C$$
$$6 \text{ electors: } \quad B \rightarrow C \rightarrow A$$

By adopting Run-off voting, the alternative C, which has the fewest votes, is eliminated. The three electors that voted for C now vote for A, and so, eventually, the winner is A, with 7 votes against 6 for B. Suppose that, at the last minute, two more electors vote with rankings

$$2 \text{ electors: } \quad C \rightarrow A \rightarrow B$$

Even if they don't vote for A, they do, in any case, prefer A to B. The effect of the two new electors is to raise the number of votes for C, so that the least voted-for alternative becomes A, who is therefore eliminated. Between B and C, the winner is now B. One might tell the new electors: don't vote, your vote will hand victory to the person whom you like the least!

7.8   Choice and Rank Monotonicity

If a collective ranking has already been derived by a social profile, we are interested to know what can happen if some preferences are changed. This aspect has little practical relevance in political elections, because it is never permitted for it to change the expressed votes. It is true that, with Run-off voting, electors vote more than once and so it may happen that some electors change their minds, but the second ballot takes place in a different political context and is not directly comparable to the first ballot.

Also, in this case, we are interested to see whether certain situations, which should happen according to common sense, actually may not occur. In particular, if a candidate has a certain place in the collective ranking and, in some individual rankings, that position is improved, we expect the candidate's position in the collective ranking not to become worse. This requirement is called the *Choice monotonicity criterion*.

Analogously, if a candidate turns out to be the winner and, in some individual rankings, that candidate's position is improved, we would not only expect that candidate to remain the winner but also that the rankings of the other candidates would not change. This requirement is called the *Rank monotonicity criterion*.

Well, a theorem by Balinski, Jennings and Laraki [11] states that *there is no unanimous, impartial rule of voting that is both choice monotone and rank monotone*.

Chapter 8
Majority Judgement

In recent years, a novel way of approaching the problem of collective choice has been proposed by Balinski and Laraki.[1] As suggested by the authors themselves, this method could be summarized in the motto: *Judge, Don't Vote!*

Before describing their approach, it is useful to quote the following remarks by the authors [15]: "What is amazing about the theory of social choice is that the basic model has not changed over seven centuries. Comparing candidates has steadfastly remained the paradigm of voting. And yet, both common sense and practice show that voters and judges do not formulate their opinions as rank orders. Rank orders are grossly insufficient expressions of opinion, because a candidate who is second (or in any other place) of an input may be held in high esteem by one voter but in very low esteem by another."

This criticism of the traditional method of voting is not new. We recall the criticism already expressed in 1925 by Walter Lippman in his essay 'The Phantom Public', which we quoted in the Introduction [59].

Balinski and Laraki, with their method known as Majority judgement, try to overcome the difficulties of the traditional method of voting. There are two main aspects in which their technique differs from the usual methods. As is made clear in the above citation, rankings among alternatives are avoided. Rankings were the starting point for Condorcet's and Borda's methods, as well as for Arrow's theory. Contrastingly, in Majority judgement, every elector expresses a judgement on the alternatives, directly and without making comparisons between alternatives. The judgement is not and cannot be of a numerical type, because, as we have previously remarked, numbers do not have the same meaning for all electors and, for this simple fact and also because number scales are rarely uniform, numbers cannot be used to

---

[1] Michel Balinski (1933–2019) was an applied mathematician who contributed important results to Operations Research and Voting Theory. In 1982, he and H.P. Young published the fundamental monography 'Fair representation: meeting the ideal of one man, one vote', a work that has been quoted by the U.S. Supreme Court. Rida Laraki (1974) worked with Balinski on the Majority judgement method. He is currently Director of Research CNRS in computer science at LAMSADE, Paris.

© Springer Nature Switzerland AG 2020
P. Serafini, *Mathematics to the Rescue of Democracy*,
https://doi.org/10.1007/978-3-030-38368-8_8

make sums and averages. The judgement instead uses the current language in such a way that the meaning of the words used is, to the greatest degree possible, the same for all electors.

The second aspect, strictly related to the first, is that, given the impossibility of making sums or averages, one employs the median to establish who gets the maximum consensus. To apply the median, it is only required that the data be ordered (from the best to the worst, or vice versa). It is not necessary that the data be numbers. The median is that data for which half of the data are better than the median and half of the data are worse. The concept of a median is very important in statistics and conveys information that is quite different from that conveyed by the mean (which can be applied only to numbers) in the description of a phenomenon. For instance, if we say that the average survival time following a diagnosis of cancer is one year, the patient can expect to live for more or less one year. However, if one year is the median, this means that the probability of living more than one year is 50%, and this implies that one could live several years more.

8.1   The Median and the Majority Grade

The median was introduced by Francis Galton [41] in consideration of juries that have to decide how to allocate money, e.g., for a project or an insurance pay-off: "...that conclusion is clearly *not* the *average* of all the estimates, which would give a voting power to 'cranks' in proportion to their crankiness. I wish to point out that the estimate to which least objection can be raised is the *middlemost* estimate, the number of votes that it is too high being exactly balanced by the number of votes that it is too low. Every other estimate is condemned by a majority of voters as being either too high or too low, the middlemost alone escaping this condemnation".

To better explain how the cranks cannot influence the final decision if the median is employed, let us take, as an example, the numbers 8, 8, 7, 5 and 4. The median is 7 (two numbers are smaller and two are bigger), and the mean is 6.4 (the sum divided by 5).

If the two smallest values (5 and 4) are changed into other values, ones that are still smaller than 7, the median remains 7, and, analogously, if the two biggest numbers (8 and 8) are changed into other values, ones that are still bigger than 7, the median does not change. This does not happen with the mean.

In statistics, the median is considered a robust indicator, especially in the presence of the so-called outliers, that is, anomalous or abnormal values that could result from errors, and therefore are better excluded from the 'normal' data. For instance, if a brilliant student takes a tumble in an exam (and the same reasoning applies for a lousy student who unexpectedly passes an exam with flying colors), this outcome has the same influence on the median of a score slightly below the median (or above, in the other case).

This observation also helps in understanding that, in a case in which a judge has to assign points to candidates, excessively increasing the points to the favorite candidate or excessively decreasing the points to the competitors of the favorite candidate, the so-called strategic vote, has no effect.

Therefore, the median represents the judgement of maximum consensus, or, alternatively, of least objection, and so it can be used as a value of collective choice. Now, we are going to use the term *grade* for the judgements that the electors assign to the candidates. So, in the previous examples, the indicated numbers are grades.

If the number of electors is odd, the median is univocally determined. The grades are ordered from the best to the worst, with the grade that is exactly in the middle being the median. If, for example, we have the five values 7, 1, 4, 8 and 5, we first sort them as 8, 7, 5, 4 and 1, and then we take 5, which lies in the middle, as the median. A grade of 5 represents the judgement of maximum consensus among the five electors. There is an exact balance between those that have expressed a more favorable judgement (those who voted 8 or 7) and those that have expressed a less favorable judgement (those who voted 4 or 1). If we had chosen 6 as a collective preference grade, we would have an imbalance between the more favorable judgements (8 or 7) and the less favorable ones (5, 4 or 1).

If the number of electors is even, there are two median values[2] and we have to choose between the two. If, for instance, we have grades corresponding to the numbers 8, 7, 5 and 4, the choice of the collective grade is between the medians 7 and 5. Balinski and Laraki recommend taking the less favorable of the two grades for the following reason: suppose that an alternative gets the two grades 8 and 2 and another one gets the grades 6 and 4. Note that they have the same mean. Should we choose according to the two highest grades 8 and 6 or to the two lowest grades 2 and 4? It seems sensible to choose according to the lowest grades, because choosing an alternative that has received a low judgement (the one with values (8, 2)) presents a greater degree of risk than choosing the other one.

Given this particular choice, it is convenient to use a different term to denote this value. The term chosen by Balinski and Laraki is *majority grade*. To sum up: the majority grade is the median when the number of data is odd and it is the less favorable of the two medians when the number of data is even.

It is useful once more to reaffirm the type of information that is expressed through the majority grade as applied to an electoral contest: *the majority grade is that grade for which there exists an absolute majority of electors that are against a less favorable judgement and also an absolute majority or a parity of electors that is against a more favorable judgement.*

Looking at the previous example with the five grades 8, 7, 5, 4 and 1, there are three electors who are against a judgement less favorable than 5 (casting votes 8, 7 and 5), hence an absolute majority. At the same time, there are also three electors who are against a judgement more favorable than 5 (casting votes 5, 4 and 1), again an absolute majority.

---

[2]and also all numbers, not necessarily integers, that lie between these two, if we also admit different values with respect to those that have been assigned. This possibility exists only with rational or real numbers. Since we are going to use non-numerical data we consider here only the two given numbers and not the intermediate ones.

If we look at the example of the four electors who assign the grades 8, 7, 5 and 4, with 5 being the majority grade, there is also an absolute majority of electors against a less favorable judgement (casting votes of 8, 7 and 5). However, the electors against a more favorable judgement are exactly half (casting votes 5 and 4). The asymmetry between the two cases is due to the fact that, when the number of electors is even, we have decided to take the lower value between the medians as the majority grade. So, the choice of the lower grade can also be justified by this consideration: the majority against a less favorable judgement must be absolute, whereas, against a more favorable judgement, we may also admit parity.

These considerations are also valid when there are many equal grades, a normal circumstance when there are many electors and few judgement grades, as it happens in political elections. For instance, if the grades received by an alternative are 10, 9, 5, 5, 5, 4, and 3, the majority against a judgement less favorable than the majority grade 5, is five electors out of seven, and likewise for a more favorable judgement.

The majority grade can also be defined as follows. There are sporting competitions in which the best and worst scores are discarded and the collective score is based only on the remaining values. This choice is motivated by the idea that the extreme judgements could turn out to be anomalous (for instance, they could result from strategic voting), and so it is not worth taking them into consideration. Recently, the Fédération International de Natation (FINA) has introduced new rules for the scores in diving competitions [38]. If there are seven judges, the two highest scores and the two lowest scores are discarded. So, there remain only three judgements for the collective score. If they were to have gone a little bit further by discarding the three highest scores and the three lowest scores, there would remain but one judgement, clearly the majority grade. Hence, we may think of obtaining the majority grade by eliminating, in turn, the best judgement and then the worst judgement (among the remaining ones) until there is only one judgement left, which is the exact majority grade.

In conclusion, an alternative is evaluated by looking at its majority grade. Of course, it may happen that two alternatives receive the same majority grade, and thus we must differentiate between them. This is very simple and is done by dropping the majority grade judgements from both alternatives and computing the majority grade again. This second majority grade represents a second value of maximum consensus. If the second majority grades are also equal, then the procedure is repeated, that is, the second majority grade is dropped from both alternatives, and so on.

Then, we may think of sorting all grades that an alternative has received in the order of the computed majority grades, and this new ordering can be called the *majority value*. If, for instance, an alternative has received the grades 8, 8, 7, 5 and 4 from five different electors, its majority value is (7, 5, 8, 4 ,8), which derives from this sequence of operations (the majority grades that are dropped one at a time are in boldface): (8, 8, **7**, 5, 4) → (8, 8, **5**, 4) → (8, **8**, 4) → (8, **4**) → (**8**). If another alternative receives the grades 9, 8, 7, 6 and 3, then its majority value is (7, 6, 8, 3, 9).

We may also sum up the way that the majority value is computed as: first, all grades are sorted from the best to the worst, and then, starting from the majority grade, one takes the grades from both sides, alternating between right and left, moving farther from the majority grade.

Then, two alternatives are compared by comparing the two majority values in the so-called lexicographic way: one looks at the first grades in the majority values and whoever has the better grade is the better one; if the first grades are equal, one looks at the second grade and whoever has the better second grade is the better one; if the second grades are also equal, one looks at a the third grade, and so on. So, between the two evaluations of the previous paragraph, the second alternative prevails, because the two first grades in the two majority values are equal (7), but the second grades are different and 6 prevails over 5.

A draw is possible only if two alternatives have the same number of grades for each grade, a case that is almost impossible with many electors, whilst, with few electors (like the juries in musical contests or sporting competitions), the probability is low, but not so low as to practically exclude the circumstance of a draw.

8.2   Grade Language

For the sake of simplicity, we have shown examples that use numbers. However, we said at the beginning that numerical evaluations are to be avoided. Hence, in order to employ the Majority judgement in an electoral context, we first have to decide which grades, based on natural language, the electors have to use. Balinski and Laraki [15] propose the following seven grades in decreasing order:

Outstanding - Excellent - Very good - Good - Acceptable - Poor - to Reject

Of course, other types of grades can be employed, even in different numbers. However, the tests carried out by Balinski and Laraki recommend using a limited number of grades. As to the fact of having seven (more or less) degrees of judgement (the grades), one can read a famous and often cited psychology paper by Miller, 'The magical number seven, plus or minus two: Some limits on our capacity for processing information' [66].

Each elector must assign one of these seven grades to each alternative. Concretely, he/she could do it by marking a cross in correspondence to an alternative and the grade for that alternative. As an example, with four alternatives A, B, C and D, an electoral ballot could be filled out as shown in Fig. 8.1, in which we have denoted the seven grades only with initials (the cross could be missing for some alternative in the improbable, but conceivable, case of no judgement; also, having more than one cross for an alternative could result in an invalid vote for that alternative).[3]

---

[3] We add one observation concerning the possibility of identifying a voter from the expressed grades. The malpractice of instructing people as to how to vote and then checking whether they actually did so is, unfortunately, present in some Italian regions. Since there are many ways of filling out a ballot of grades, this worry is justifiable. For instance, with eight alternatives and seven grades, this number is 5,764,801, sufficiently high to identify each elector in a large area. Clearly, the malpractice also needs to steer the outcome and not all grades can serve the purpose. At most, two

	O	E	V	G	A	P	R
A			X				
B						X	
C		X					
D				X			

Fig. 8.1 Possible electoral ballot for Majority judgement

8.3 Collective Ranking

From all ballots, one counts how many Outstanding grades (O) have been assigned to each alternative, how many Excellent grades (E), and so on. Then, the majority grade is computed. For instance, out of 100 electors, an alternative might receive 20 Outstanding grades, 13 Excellent, 10 Very good, 17 Good, 10 Acceptable, 12 Poor and 18 to Reject. The majority grade is Good. Indeed, by sorting all of the grades in decreasing order, the 51st grade is Good (100 is even, and we have to take the lower grade between the 50th and the 51st; in this case, they are equal and there are no differences).

Moreover, we may say that 43 electors have assigned a grade higher than Good, so that $43 + 17 = 60$ electors, an absolute majority, are against a judgement that is lower than Good. Symmetrically, 40 electors have assigned a grade lower than Good, so that $40 + 17 = 57$ electors, an absolute majority, are against a judgement higher than Good. The grade Good therefore represents the maximum consensus for the alternative.

To obtain the majority values necessary to rank all alternatives, we should repeat the operation of dropping the majority grades one at a time and list the grades thus obtained, as previously explained. With a large electorate, this would be a very lengthy operation, but, luckily, we can operate more quickly. It is clear that, by dropping all Good grades one at a time, at a certain point, one will come across a majority grade of either Very good or Acceptable, that is, the two grades adjacent to Good. Which one will appear first as a majority grade? If the number of electors that have given a better grade than Good is larger than the number of those that have given a worse grade than Good, then Very good will appear first. In the opposite case (less than or equal to), Acceptable will appear first. Hence, in the example, Very good will appear first (43 vs. 40).

Each time the majority grade has more electors that have expressed a better grade than those who have expressed a worse grade, the majority grade is marked with a +. If the opposite happens, it is marked with a −. So, if the numbers of better and worse grades are equal, one marks the majority grade with a − and not with a +,

---

or three grades could be used. Yet, the number of different ballots is high. In any case, one way to deter this criminal behavior does exist. It is enough to materially detach the alternatives in an electoral ballot at the very time of voting. Practically, this is feasible.

because we have decided to take the lower median when we have an even number of judgements. In the example, we have Good+.

To obtain a ranking, we also need to take into account how many better and worse grades there are. In conclusion, we have to consider:

- the majority grade;
- whether it is marked with + or −;
- the number of better grades;
- the number of worse grades.

The number of grades can also be indicated as percentages. For instance, the alternative in the example has a majority value that can be summarized by the three pieces of data $(43, B+, 40)$, where the first is the percentage of grades better than the majority grade, the second is the majority grade marked with + or −, and the third is the percentage of grades worse than the majority grade. This kind of succinct information in regard to the majority value is called the *majority gauge*.

To rank the alternatives, we first order according to the majority grade, and, majority grades being equal, whichever one is marked with + prevails over those marked −. If two alternatives have the same mark +, one looks at the number of better grades, and whichever one has the greater number of better grades prevails. If they have the same mark −, one looks at the number of worse grades, and whichever one has fewer worse grades prevails. With a large electorate, the probability of a draw is negligile. If a draw does occur, then we resort to the majority values and proceed as previously explained.

In Fig. 8.2 we show an example of the judgements of 100 electors over three alternatives A, B and C. Each bar represents how the grades for that alternative are assigned. On each bar, the percentages of the various grades are proportionally displayed (O = Outstanding, E = Excellent, etc.). The bar at the top represents the alternative A, whose grades we have previously indicated: $O = 20, E = 13, V = 10, G = 17, A = 10, P = 12, R = 18$. The alternative B (central bar) is marked by the following grade percentages: $O = 10, E = 15, V = 30, G = 10, A = 10, P = 15, R = 10$. The alternative C (bar at the bottom) is marked by: $O = 5, E = 22, V = 25, G = 5, A = 6, P = 10, R = 27$.

The vertical segment crossing the three bars represents the majority grade. The dotted segments identify the grades as being either better or worse than the majority grade and the numbers indicate their respective quantities.

As can be seen from Fig. 8.2, the majority grade for the alternative A is Good+ (this has already been computed above), whereas for B and C is Very good−. The mark − is due to the fact that the grades that are worse than Very good are greater in number than the better grades. So, B and C are better than A. For the ranking, we have to discriminate between B and C. Since they have both Very good−, one must look at how many a worse grades there are. Since B has fewer worse grades than C, B prevails over C. In conclusion, the final ranking is $B \Rightarrow C \Rightarrow A$. The majority gauges for this example are:

$$A \to (43, G+, 40), \quad B \to (25, V-, 45), \quad C \to (27, V-, 48)$$

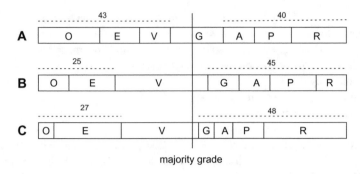

Fig. 8.2 Majority grades for three alternatives

This new method of ranking alternatives, the Majority judgement, has begun to be employed in some specific cases, like wine competitions, musical contests, figure skating. Maybe its application to general political elections seems utopian, given the very large number of electors. The main objection is that filling in the electoral ballot requires a certain degree of effort from the voters. However, the effort does not seem to be higher than that of providing a ranking, as is required in Condorcet's and Borda's methods.

We note that, if there are only two grades, Majority judgement corresponds exactly to Approval voting. The better grade is for the approved candidates and the worse grade is for the non-approved ones. The obtained rankings are the same. So, Approval voting can be seen as a very particular case of Majority judgement. Having a wider spectrum of possible grades makes Majority judgement a more refined way of voting.

Majority judgement could also be used in a parliamentary system like Italy's or Germany's to elect the President of the Republic through the parliament (in Italy, a modification of the Constitution would be required to change the current method of electing the President). Given the presence of a strong strategic component in the vote of the parliament members, grouped into various political positions, this could be a very interesting test to evaluate the robustness of Majority judgement with respect to the Non-manipulability issue.

It is, anyway, true that an increasingly urgent need is perceived to formulate electoral systems that are free from paradoxes, anomalies induced by strategic voting and other defects that arise from a grossly insufficient method of expressing a political wish, like Single vote. When the outcome of an election risks being in disharmony with the 'hidden' preference of the electorate, the danger is likely greater than that of suppressing the free vote. In the latter case, whoever governs does not in any way represent the popular will, but at least it is an evident fact. In the former case, the government is legitimate, but does not faithfully represent the popular will, and this can create such a significant degree of unrest as to lead to unpredictable outcomes.

8.4 Domination in Majority Judgement

We add some considerations that concern those cases in which, by adopting the reasoning of Majority judgement, there should be no doubts about who is the winner. Imagine a simple case with two alternatives A and B and eight electors. The grades are those that have been previously listed and are denoted here only with the initials. Once the grades are ordered for each alternative, from the best to the worst, the two alternatives are thus evaluated:

A	O	E	E	V	G	G	A	A
B	O	E	V	G	G	A	A	P

For each position in the table, a grade of A is always equal to or better than one of B. It seems natural to deduce that A must be collectively preferred to B. When such a situation arises, we say that A dominates B. More exactly, A *dominates B if all (ordered) grades of A are better than or equal to the (ordered) grades of B and at least one is better*.

To avoid misunderstanding, we should note that the grades listed one above the other *do not necessarily correspond to the same elector*. The above table could result from the following individual profiles, in which, contrastingly, each column refers to a particular elector:

electors	1	2	3	4	5	6	7	8
A	A	G	O	G	E	V	E	A
B	O	E	V	G	G	A	A	P

From this table, we see that A prevails over B five times (electors 3, 5, 6, 7 and 8), B prevails over A two times (electors 1 and 2) and A and B receive the same grade once (elector 4). So, we also have $A \Rightarrow B$ according to the Majority criterion. This second table, however, does not play any role in Majority judgement, which, as we have already stressed, *does not* work in comparisons between alternatives and considers only the first table.

The profiles in the second table are not the only ones that can produce the first table. It is enough to assign each grade of the first table to an arbitrary elector for us to obtain a different collective profile (having many equal grades reduces the number of different profiles that we can obtain).

If, instead of having the single grades, we have the grade percentages, the domination concept is translated as: A dominates B if its percentage of the highest grade is greater than or equal to that of B, and also if the sum of the two highest grades

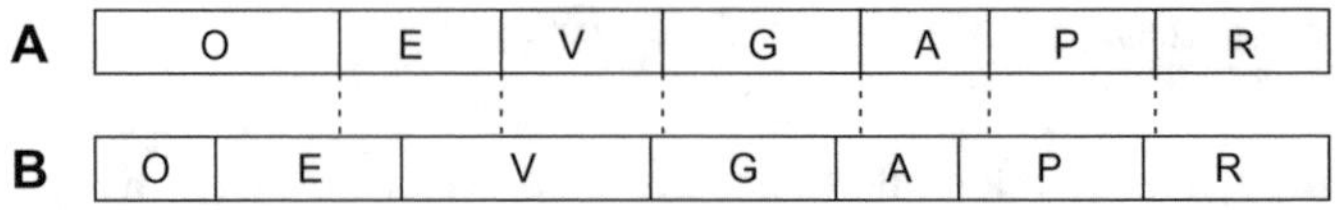

A	O	E	V	G	A	P	R

B	O	E	V	G	A	P	R

Fig. 8.3 Dominance of A over B

of A is equal to or greater than the same sum for B, and also if the sum of the three highest grades for A is equal to or greater than the same sum for B, and so on up to the second-to-last grade, with at least one case being strictly greater. If we visualize the percentages as bars (as we have previously done in Fig. 8.2), A dominates B if every one of A's grades ends after (or at) the same grade as B (with 'after' occurring at least once), as it happens in Fig. 8.3. For the three bars in Fig. 8.2, there is no dominance between any pair of alternatives.

8.5   The 2012 French Presidential Elections

Before comparing Majority judgement to other electoral methods, it is certainly interesting to observe which result can be obtained in a real electoral contest. Even if the method has never been really employed in political elections, a simulation was, however, carried out some days before the first ballot of the French Presidential election of April 22, 2012, on a group of 993 electors. These electors were subsequently reduced to 773 in order to have more statistical coherence with the real election outcome. Each one of these electors was asked to vote for the ten candidates, both with Single vote and with Majority judgement. For the top five candidates, they were also asked to vote with Condorcet's method. Moreover, an Approval voting was carried out with other electors.

As far as Majority judgement is concerned, each elector was asked to assign to each candidate one grade out of seven by using an electoral ballot like the one in Fig. 8.1. In French, the seven grades were: Excellent, Très bien, Bien, Assez bien, Passable, Insuffisant, A rejeter. In Table 8.1, we report from [13] the percentages (with respect to the 773 electors of the simulation) of the assigned grades. The grades are denoted by the English initials of the seven grades that we have already presented. Table 8.1 represents numerically what was graphically shown in Fig. 8.2 (whose data, of course, was not related to the French elections).

The results of the simulation are reported in Table 8.2 that we reproduce from [13, 15]. The first column reports the names of the ten candidates ordered according to the Majority judgement ranking. In the second column (MJ), the majority gauges are reported. These have been computed as explained in the previous section. For instance, for Hollande, the sum of the percentages for the grades Outstanding, Excellent and Very good gives $12.48 + 16.15 + 16.42 = 45.05\%$. So, the percentage of electors that judged him to be more than Good was 45.05%, whereas the percentage that judged him to be at least Good was $45.05 + 11.67 = 56.72\%$, an absolute majority. Moreover, the sum of the percentages for the grades Acceptable, Poor and

Table 8.1 Majority judgement percentages for the 2012 French Presidential election

	O	E	V	G	A	P	R
Hollande	12.48	16.15	16.42	11.67	14.79	14.25	14.24
Bayrou	2.58	9.77	21.71	25.23	20.08	11.94	8.69
Sarkozy	9.63	12.35	16.28	10.99	11.13	7.87	31.75
Mélenchon	5.43	9.50	12.89	14.65	17.10	15.06	25.37
Dupont-Aignan	0.54	2.58	5.97	11.26	20.22	25.51	33.92
Joly	0.81	2.99	6.51	11.80	14.66	24.70	38.53
Poutou	0.14	1.36	4.48	7.73	12.48	28.08	45.73
Le Pen	5.97	7.33	9.50	9.36	13.97	6.24	47.63
Arthaud	0.00	1.36	3.80	6.51	13.16	25.24	49.93
Cheminade	0.41	0.81	2.44	5.83	11.67	26.87	51.97

Table 8.2 Majority judgement (MJ), Single vote (SV) and Approval voting (AV) for the French Presidential election 2012

	MJ	SV	AV
1. Hollande	(45.05, G+, 43.28)	(1) 28.63	(1) 49.44
2. Bayrou	(34.06, G−, 40.71)	(5) 9.09	(3) 39.20
3. Sarkozy	(49.25, A+, 39.62)	(2) 27.27	(2) 40.47
4. Mélenchon	(42.47, A+, 40.43)	(4) 11.00	(4) 39.07
5. Dupont-Aignan	(40.57, P+, 33.92)	(7) 1.49	(8) 10.69
6. Joly	(36.77, P−, 38.53)	(6) 2.31	(6) 26.69
7. Poutou	(26.19, P−, 45.73)	(8) 1.22	(7) 13.28
8. Le Pen	(46.13, P−, 47.63)	(3) 17.91	(5) 27.43
9. Arthaud	(24.83, P−, 49.93)	(9) 0.68	(9) 8.35
10. Cheminade	(48.03, R+, —)	(10) 0.41	(10) 3.23

to Reject gives $14.79 + 14.25 + 14.24 = 43.28\%$. So, the percentage of electors that judged him to be less than Good were 43.28%, whereas the percentage of those that judged him to be at most Good were $43.28 + 11.67 = 54.95\%$, an absolute majority. Hence, his majority grade is Good. The data 45.05% (better than Good) and 43.28% (worse than Good) are reported in the majority gauge. Since the number of electors who judged him to be more than Good is higher than that of those who judged him less than Good, the majority grade is marked with +. For the other candidates, the procedure is the same.

In the third column (SV) of Table 8.2, the percentages of Single vote and the ranking number with respect to Single vote are reported. In the last column (AV), the percentages of Approval voting and the ranking number with respect to Approval voting are reported. In this last case, the percentages are the number of marks with respect to all electors and the sum, of course, is larger than 100%.

We recall that, in the second ballot on May 6, 2012, Hollande won by a narrow margin against Sarkozy. In the first ballot, Hollande turned out to be the most voted-for and the actual results agree well with the data in column SV. This should be no surprise, given the decision to reduce the number of electors chosen for the simulation from 993 to 773, by excluding those who were too discordant with respect to the actual data. Through this exclusion, it is expected that the results for the other election methods will also be more realistic.

With Single vote, Marine Le Pen came in third, a remarkable personal success. The percentage that Hollande got in Single vote is affected by the competition from Bayrou on the right and Mélenchon on the left. By allowing the electors to increase the information with Approval voting, the ranking changes. The competition among similar candidates disappears: Hollande, Sarkozy, Bayrou and Mélenchon increase their percentages by a significant extent and Bayrou moves from fifth to third place. Le Pen, who suffers from a very polarized consensus in regard to her person, is overtaken by Bayrou and Mélenchon and drops to fifth place.

Lastly, by also adding the information provided by the Majority judgement, the attraction-repulsion effect of a candidate like Le Pen emerges clearly. Her position further drops to eighth place. Moreover, Bayrou becomes second, replacing Sarkozy. As is clearly displayed in the table, Single vote hides a large part of the opinion of the electors.

It is also quite interesting to see whether there are candidates who dominate other candidates. For this purpose, we have to build a table in which we indicate the cumulative sums of the percentages. From Table 8.1, we form Table 8.3, in which the first column reports the percentages of Outstanding, the second column the sums of the percentages of Outstanding and Excellent, the third column the sum of the percentages of Outstanding, Excellent and Very good, and so on. One candidate dominates another one if all of his/her figures in this table are bigger than those of the other candidate (some may be equal). From Table 8.3, it can be seen that there are

Table 8.3 Cumulative sums of the Majority judgement percentages for the 2012 French Presidential election

	O	O+E	O+E+V	O+$\cdots$+G	O+$\cdots$+A	O+$\cdots$+P
Hollande	12.48	28.63	45.05	56.72	71.51	85.76
Bayrou	2.58	12.35	34.06	59.29	79.37	91.31
Sarkozy	9.63	21.98	38.26	49.25	60.38	68.25
Mélenchon	5.43	14.93	27.82	42.47	59.57	74.63
Dupont-Aignan	0.54	3.12	9.09	20.35	40.57	66.08
Joly	0.81	3.80	10.31	22.11	36.77	61.47
Poutou	0.14	1.50	5.98	13.71	26.19	54.27
Le Pen	5.97	13.30	22.80	32.16	46.13	52.37
Arthaud	0.00	1.36	5.16	11.67	24.83	50.07
Cheminade	0.41	1.22	3.66	9.49	21.16	48.03

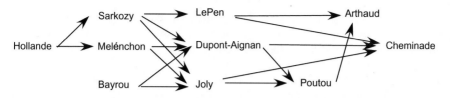

Fig. 8.4 Domination graph for the 2012 French Presidential election

Table 8.4 The Condorcet table and the Borda count for the 2012 French Presidential election

	Hollande	Bayrou	Sarkozy	Mélenchon	Le Pen	Borda count
1. Hollande	—	51.6	53.9	68.5	64.1	238.1
2. Bayrou	48.4	—	56.5	59.4	70.5	234.8
3. Sarkozy	46.1	43.5	—	50.5	65.7	205.8
4. Mélenchon	31.5	40.6	49.5	—	59.7	181.3
5. Le Pen	35.9	29.5	34.3	40.3	—	140.0

many cases of domination. For instance, Hollande dominates all candidates except Bayrou. In Fig. 8.4, we indicate the dominance between two candidates with an arrow (the graph is the so-called transitive reduction of the domination graph, that is, if A dominates C, but A also dominates B and B dominates C, the domination of A over C is not indicated, because it is implied by the other two).

The simulation results for Condorcet's method are reported in Table 8.4. The figures represent the percentages of voters who prefer the candidate in the row to the one in the column. It turns out that Hollande is the Condorcet winner and the ranking agrees with the one produced by Majority judgement. From the table, we may also compute the Borda count by summing the numbers in each row (the last column of Table 8.4). Borda's method also reproduces that ranking.

It is interesting to see that three methods, all of which use a great deal of information, yield a ranking that strongly disagrees with the one provided by the current voting system, i.e., Single vote, and even that provided by Approval voting.

8.6 Majority Judgement and Condorcet

We have seen that, in the case of the French Presidential election, both Majority judgement and Condorcet's method produce the same result. We may wonder whether the two methods always agree, in the sense that, each time there exists a Condorcet winner, this is also first in the ranking of Majority judgement. The answer is negative. Vice versa, we may also ask whether, when one alternative dominates another one according to Majority judgement, that same alternative is also preferred according to Condorcet. The answer is again negative.

Before comparing the two methods, we have to note that they use different information gathered from the elector and so the comparison has to be made with caution. The information that Majority judgement requires from the elector is partly superior and partly inferior to that required by Condorcet's method. Surely, if alternative A receives a better grade from an elector than alternative B, we may conclude that A is preferred to B for that elector. For instance, if we look at the electoral ballot in Fig. 8.1, we may deduce that $C \to A \to D \to B$ for that particular vote. But, in general, two alternatives may receive the same grade, and so we might think of reporting them in a Condorcet table as having an indifference relationship. In fact, we don't know if the elector, even if he/she assigns the same grade, prefers one to the other. So, the information provided by the grades is partly inferior to that provided by a ranking.

From the other side, given a preference $A \to B$, we don't know what grades underlie that preference. For instance, we don't know whether $A \to B$ comes from the grades Outstanding and Excellent (respectively for A and B) or from the grades Outstanding and Poor, or even from Poor and to Reject. It is clear that the three situations are very different from each other. Hence, the information given by the grades is partly superior to that from the ranking.

Moreover, a surprising fact may happen. Suppose that the electors have in their minds some judgements concerning the alternatives and, fruiiom these judgements, a collective profile of preferences is derived. Let us consider the simple case of two alternatives the preferences for which directly give a majority winner. For instance, 100 electors express the grades Good, Acceptable and Poor, as in the table (the example is reproduced from [15]):

	30	10	10	25	25
A	G	G	A	A	P
B	A	P	G	P	G

Hence, A is preferred to B by 65 electors (first, second and fourth columns) and we have $A \Rightarrow B$ in what we normally would consider to be a clear victory. Now, we transform the profile as required by Majority judgement (which, we repeat once more, does not operate according to comparisons) and get

	G	A	P
A	40	35	25
B	35	30	35

According to Majority judgement, A is preferred to B, because its majority grade is Acceptable+, while the majority grade of B is Acceptable− (we recall that, when the number of higher grades is equal to the number of lower grades, a − is assigned).

Moreover, A dominates B (A has more Good grades than B and also a higher number of Good and Acceptable combined). Therefore, for this example, the two methods agree. Now let us suppose that the 100 electors have other different judgements, as in the following table:

	5	35	35	25
A	G	G	A	P
B	A	P	G	A

from which it turns out that B is preferred to A by 60 electors (third and fourth columns) and so $B \Rightarrow A$. This second profile produces a different outcome. But, if we transform this profile according to Majority judgement, we obtain the following table:

	G	A	P
A	40	35	25
B	35	30	35

which is exactly equal to the one obtained with the other profile! So, according to Majority judgement, A should be the winner, and not B. We recall that A even dominates B in Majority judgement, but apparently, in this case, Majority judgement does not respect the Majority criterion.

Why does this discrepancy occur? The reason is that, with Condorcet's method, it is enough to denote the existence of a preference without saying anything about the extent of the preference, whereas Majority judgement takes account of this quantitative aspect. The fact that two different profiles give raise to the same Majority judgement can be explained by looking at the figure of those 35 electors, who, in the second profile, assign the grade Good to A and Poor to B. So, a large share of the electorate prefers A to B in a clear-cut way. However, in Condorcet's method, we see only the existence of this preference and not its size, which is therefore lost. The result of this 'cancelation' is that, with the second profile, B wins, both with Condorcet's method and with any other method that is based only on the mere existence of preferences.

Just note that the example has only two alternatives, a case in which everything should be simpler. On the contrary, we can face a dilemma just as easily with only two alternatives. If the society expresses the second profile, who should be the winner? If we decide to adopt Condorcet's criterion, which, for two alternatives, is the exact Majority criterion, then B should win, but if we adopt the Majority judgement, A should win. There is no definite answer to the dilemma. It depends on which criterion we have decided to adopt.

In our opinion, Majority judgement better represents the will of the electorate, because it makes use of more information. In order to better understand the question, let us consider more examples. In the first example, there are five electors who give the following grades to two alternatives A and B:

	1	2	3	4	5
A	A	A	A	R	R
B	G	G	P	P	P

Elector n. 3 is the only one who prefers A to B. According to the Majority criterion, B should be the winner without any doubt. If, contrastingly, we look at the majority grades, we see that A is Acceptable and B is Poor, and so, in Majority judgement, which corresponds to the maximum consensus, A is the winner. After all, for one majority, A is at least Acceptable, and for another majority (different from the previous one), B is at most Poor. Hence, still invoking a Majority criterion, but for a different data setting, we may declare A the winner. The difference is that, in the former case, the majority is in regard to direct comparisons made by single electors, whereas, in the latter case, we are dealing with a global majority.

Let us now consider again the hypothetical case of Formula One shown on Sect. 5.2 in which a driver A wins 11 races out of 20 (hence, an absolute majority) without even participating in the remaining races and another driver B is second in those 11 races and first in the remaining 9 races. We have already observed that, both according to the Majority criterion and to Condorcet's criterion, A should be the world champion, while, according to Borda (and the actual Formula One scoring system as well), B should be the champion. What does the Majority judgement produce in this case? Actually, we don't have sufficient information to apply this method, which requires assigning a grade to each driver in each race. We can, anyway, translate the order of finishing into grades, like: First, Second, Third, ..., Tenth, NC (for not classified).

Then, driver A is assigned the grade of First eleven times and the grade of NC nine times. A's majority grade is thus First. Driver B is assigned the grade of Second eleven times and the grade of First nine times, so B's majority grade is Second. Thus, also according to the Majority judgement, A should be the winner. It is easy to see that, whoever gets the top grades for the majority of electors, will be the winner both with Condorcet and with Majority judgement. The difference with the previous example consists in this exact fact: here, the majority of better grades are the top grades and, in the previous example, B prevailed over A, but not always with the top grades.

Summing up, if we have only two alternatives and only the preference between the two alternatives for each elector and we interpret this preference as two judgement grades, one positive and the other negative, then the majority winner is also the Condorcet winner and the Majority judgement winner.

By having more articulate preferences, the situation becomes different. Let us recall the example proposed on Sect. 6.2 and written here again for convenience's sake:

$$100 \text{ electors: } A \to B \to C$$
$$200 \text{ electors: } A \to C \to B$$
$$199 \text{ electors: } C \to B \to A$$
$$201 \text{ electors: } B \to C \to A$$

If we want to translate these preferences into grades, we may think (although it is not the only conceivable method and the outcome can be different depending on how we implement it) of doing it in the simplest possible way (or, more accurately, in the most ignorant way, since we do not have the sufficient amount of information to differentiate them in some other way), that is, we use the grades First, Second and Third. This way, we have the following grades for the three alternatives:

	First	Second	Third
A	300	0	400
B	201	299	200
C	199	401	100

The majority grades are: Third+ for A, Second+ for B and Second+ for C. To differentiate between B and C, we need to look at how many grades there are that are better than Second. By a small margin (201 to 199), B prevails. So, we obtain the same outcome as with Run-off voting, but a different outcome from that of Condorcet.

What is the preference mechanism aspect that is captured by the Majority judgement and produces a different result from Condorcet? We limit ourselves to the comparison between B and C. According to Condorcet, C prevails over B 399 times, while the opposite happens 301 times, so that, as we know, C wins. For Condorcet's method, the preference between C and B when they are in second and third place has the same value as when they are in first and second place. Contrastingly, for Majority judgement, the preference between the second and third positions is less important than the preference between first and second, because being second and third already contains a negative evaluation, that is different from being first and second.

8.7 Majority Judgement and the Impossible Wishes

After these considerations, which method, should we rely on? Essentially, Social choice methods can be split into methods that make comparisons between alternatives and methods that express judgements. The Majority judgement is the only one of the second type that we have presented. All others are of the first type. The debate about which method should be adopted is open and is hardly likely to be conclusive, since a perfect method does not exist.

We list only some criteria that are considered in Social choice theory. An exhaustive review would be incompatible with the format of this book. The criteria listed below are sufficiently important. Some of them have been already introduced and discussed. Taking for granted that Unanimity and Non-dictatorship must be satisfied, a choice method should also:

– determine a ranking with a winner,
– avoid that the outcome depend on irrelevant alternatives,
– keep in consideration the vote of the majority,
– avoid strategic voting and manipulations,
– elect a Condorcet winner, when this exists,
– ensure that voters always help their favorite candidates (avoiding the No show paradox).

Majority judgement guarantees the first two conditions, but does not guarantee the last two. As far as Non-manipulability is concerned, this is partially satisfied. Note, however, that all four conditions of Arrow's theorem are satisfied. This fact does not contradict the theorem, because the theorem assumes the existence of rankings, whereas Majority judgement is based on a different type of information.

Balinski and Laraki point out that avoiding the No show paradox is actually not that important, because, as we have already remarked, the conditions for the paradox to emerge hardly ever occur in practice. Anyway, their recent theorem [16] states that, if there are, at most, three grades the paradox does not occur. As far as the disagreement with Condorcet's method is concerned, Balinski and Laraki remark that the two methods agree most of the times. When they don't, it would be useful to understand which information is more important, that coming from one method or that coming from the other (see [15] pp. 501–502).

Furthermore, we must bear in mind that no method can satisfy the first, second and second-to-last requirements at the same time [12]. Also, the last two requirements cannot both be satisfied simultaneously [67].

We still have to try understand how robust Majority judgement is with respect to strategic voting. The question can be examined from two points of view. In the first place, Majority judgement yields majority grades for the alternatives and, in the second place, it produces a ranking from the grades. If, for instance, we have to choose some people from a larger group of candidates, in the very end, it is the ranking that matters.

Let us first analyze the question concerning the majority grades. Suppose there is a contest with judges who have to elect a winner, and suppose also that the judges are not impartial because they have favorite candidates, and consequently also 'disliked' candidates. If the grade assigned by a judge to the favorite candidate is higher than the final majority grade of that candidate (a quite likely circumstance), we would desire that, if that judge changes the grade for the same candidate, the effect on the final majority grade would be either null or detrimental. Hence, the judge should have no incentive to artificially improve the grade.

Symmetrically, if the grade assigned by a judge to a candidate who is not their favorite is lower than the final majority grade of that candidate (a circumstance that is less probable than the previous one, but still plausible), we would desire that, if that judge changes the grade for the same candidate, the effect on the final majority grade would be either null or meliorative. Hence, the judge should have no incentive to artificially diminish the grade.

If a method satisfies these requirements, we can consider it robust with respect to the problem of manipulability. It has been proved [12] that only Majority judgement, among all possible methods that use judgements, can satisfy both requirements (the reason is clearly the choice of the median as a majority grade).

If, contrastingly, we consider the ranking derived by the majority grades, things are less favorable. Suppose there are two candidates A and B and that, in the final ranking, A prevails over B, i.e., $A \Rightarrow B$, but there is a judge who has the opposite opinion, namely, $B \rightarrow A$. Obviously, it cannot happen that this judge alone can overturn the final ranking. If it were so, that judge would be a dictator, and we know that Majority judgement, as well as the other methods that we have presented, satisfy the criterion of Non-dictatorship.

However, we would like to know whether this judge can, in any case, use that vote to increase the final grade for B or decrease the final grade for A, or maybe both, so as to possibly switch A with B in the ranking. It can be proved [12] that there is no voting method that can prevent the judge who prefers B to A from improving the grade for B or decreasing the grade for A. This makes sense on close inspection. If such a method existed, judges would have a non-effective vote in their hands.

Hence, we need to modify the definition of manipulability by taking into account that the vote must be effective in all cases. For instance, we may require that, if the judge who prefers B to A can increase the grade for B, then it should be impossible for him to decrease the grade for A at the same time and, symmetrically, if he can decrease the grade for A, then it should be impossible for him to increase the grade for B at the same time. It can be shown that Majority judgement satisfies this milder requirement of Non-manipulability.

It may be interesting to consider again the example on Sect. 7.6, in which the effect of strategic voting in a music contest was illustrated. In that case, each judge had to give a ranking of three candidates A, B and C. Now, we suppose that the judges have three grades available: Outstanding, Good, Acceptable. For convenience's sake, we report here again the 'sincere' rankings of the four judges that we wrote on Sect. 7.6:

judge n. 1:	$A \rightarrow B \rightarrow C$	judge n. 2:	$A \rightarrow C \rightarrow B$
judge n. 3:	$B \rightarrow A \rightarrow C$	judge n. 4:	$C \rightarrow B \rightarrow A$
judge n. 5:	$B \rightarrow C \rightarrow A$		

If we interpret the three positions in the ranking as the three indicated grades, we obtain the following table of grades:

	1	2	3	4	5
A	O	O	G	A	A
B	G	A	O	G	O
C	A	G	A	O	G

from which we obtain the majority values:

$$A \to (GAOAO), \quad B \to (GGOAO), \quad C \to (GAGAO)$$

These majority values are computed as previously explained, i.e., first, the grades are ordered from the best to the worst, for instance, for B the ordered grades are (OOGGA), and then, starting from the center and alternately picking the grades to the right and to the left moving away from the center, we get the majority value (GGOAO). We leave to the reader the simple exercise of computing the majority values for A and C.

This evaluation corresponds to Good+ for B and Good− for both A and C. Between A and C, A prevails, given the majority values. Hence, the final non-manipulated ranking is $B \Rightarrow A \Rightarrow C$, as for Condorcet's method.

Now suppose, as in the previous case, that judges 1 and 3 adopt strategic voting to make their favorite candidates win. If we recompute the majority values with the new rankings translated into grades, we see that C wins, exactly as had previously happened. However, the judges now don't have to formulate rankings, they only have to assign grades. So, judge 1 has no reason to increase his/her grade for C, it is enough to decrease the one for B (increasing A's grade is not possible, as it is already at the top). Analogously, judge 3 has no reason to increase C's grade, as it is enough to decrease the grade for A. Now, the manipulated grades are as they are in the table,

	1	2	3	4	5
A	O	O	A	A	A
B	A	A	O	G	O
C	A	G	A	O	G

and lead to the following majority values:

$$A \to (AAOAO), \quad B \to (GAOAO), \quad C \to (GAGAO)$$

The new ranking is $B \Rightarrow C \Rightarrow A$. B is still the winner, in spite of strategic voting, even if by a small margin over C, which is now second.

In conclusion, it is our opinion that Majority judgement has greater merits than other methods. Adopting this method may be complicated and departing from simple methods like Single vote, which has established roots in the commons, can certainly constitute a problem. However, electors should be aware of the limitations and distortions that such a rudimental method necessarily produces.

In France, an association has been founded, called *Mieux Voter*, with the aim of disseminating and promoting Majority judgement. The reader can find pertinent information on their website at https://mieuxvoter.fr/apropos/.

8.8 Possible Future Developments

We limit ourselves to hinting at some new questions that modern technologies have introduced in regard to voting mechanisms. The existence of the internet makes it possible to vote more than once in a fixed time span through the possibility of modifying one's vote as many times as is felt necessary. We don't deal here with the issue of the security and transparency of a vote on the web, a problem that is anything but negligible. Rather, we consider the speed that the internet has introduced and that could be exploited in the voting system.

The temporary vote outcome could be made available instantly and the elector could be allowed to modify their vote, if necessary, before the polls close. Of course, this entails engaging in strategic voting. This new possibility for voting has been the object of recent studies, even if in a slightly different way [65], in which electors vote in rounds and, as soon as there is no change between two rounds, the polls are closed. With millions of electors, it would be impossible to vote in rounds, but we feel that the analysis can be carried out with the same mathematical tools.

The fundamental question is whether there will eventually be convergence toward a certain outcome and whether this outcome is 'good'. The need for convergence is obvious. A method that would continuously move back and forth the votes, thus showing electors in a constant state of dissatisfaction, is simply inconceivable. The central issue, however, concerns the quality of the possible equilibrium point (and we cannot even be sure that it is unique). We know from Game Theory that the equilibrium point is a Nash equilibrium, and it is well known that a Nash equilibrium may find all competitors in a condition worse than the one from which they started.[4]

Besides this problem, one should bear in mind what we mentioned in Sect. 7.8. There, the possibility of changing the vote was considered as a theoretical question. In this case, it would represent the essence of the method. Hence, all of the negative results that we have presented remain valid and constitute a sound objection to the idea of a continuous voting.

It might be interesting to develop these analyses for Majority judgement, which is more robust with respect to strategic voting.

---

[4]We will come back to Nash equilibria in the final chapter (Chap. 15).

Chapter 9
Legislative Territorial Representation

The Legislative power is almost always entrusted to one or two assemblies whose members represent the nation. The representation is twofold. From one side, the various parts of the national territory must be represented and, on the other side, the various political leanings that citizens express must also be represented. Since we have to satisfy two different requirements at the same time, the problem of achieving a fair representation is by no means simple. The solutions adopted in various countries invariably present critical aspects and, in some cases, like in Italy, anomalies may occur (see Sect. 12.3). In this chapter we deal with the territorial representation and, in Chap. 11, we deal with the political representation.

9.1 General Criteria

We first separately present the problem of territorial representation, which, in and of itself, is not trivial. Preliminarily, the whole nation is divided into a certain number of districts.[1]

In some countries, like the United Kingdom, each district sends a unique representative to the House. In this way, the number of House members is equal to the number of districts and the problem of how many representatives should be assigned to a district does not exist.

Having escaped one problem, two more problems arise. One concerns the definition of a district. The way in which to draw a district map is a very critical issue, because one may change an electoral outcome with a few pen strokes. We will come back to this crucial and complex problems in Chap. 13.

---

[1] Subsequently, we will deal with only one of the two chambers that are usually present in many countries, the so-called Lower House. The Upper House follows a particular set of rules in many cases. In Italy, the Upper House, namely, the Senate, follows a logic similar to that of the Lower House, the Chamber of Representatives (Camera dei Deputati). In any case, by 'House' and 'Parliament', we are always referring here to the Lower House.

© Springer Nature Switzerland AG 2020
P. Serafini, *Mathematics to the Rescue of Democracy*,
https://doi.org/10.1007/978-3-030-38368-8_9

The other problem concerns the distortion that arises between the vote percentages obtained by the various parties and the respective percentages of seats gained. For instance, in the 2005 elections in the United Kingdom the Labour Party obtained 57% of seats in the face of only 36% of votes. In the 2010 elections, the vote percentages for the Conservatives, Labour and Liberal Democrats were 36%, 29% and 23%, respectively, but the seat percentages were 47, 39.7, and 9% [96].

Furthermore, the candidate that takes a seat in the parliament is chosen according to the Simple majority criterion, i.e., the winner is the one who gets more votes, and we discussed at length in Chap. 6 how wrong this method of choice can be.

A pure majority system, like the British one, corresponds, in our opinion, to a 'primitive' system of representing a nation. In origin, it makes sense to send a person to the House who people believe is the best representative of the interests of that particular territory. However, with the progress of a nation, the territorial interests must be reconciled within a superior global vision of the entire nation's interests. To this end worldviews, ideologies and parties emerge and these have to be represented in a coherent and predominant way. The pure majority system warps this representation and, if, in addition, the government formation is entrusted to the parliament and not to a direct vote, the misrepresentation occurs at both the legislative and executive levels.

So, the larger the number of seats per district, the more faithful the political representation at the national level. At the limits, the ideal political representation would correspond to one district that is as large as the whole nation. But, in this way, we would loose the territorial representation. Then, we need to find the right compromise in which the districts are sufficiently small for there to be good representation of the territory, but also sufficiently large to allow for an adequate number of seats, and therefore a better political representation.

In order to decide how many seats should be globally assigned to the House and how they should be divided among the districts, one always refers to the present population, and not to the people who have the right to vote or to the actual number of voters, with the idea that all people must be represented, even if they are not of legal age or they abstain from voting. The number of global seats is usually fixed beforehand and is linked to the population data for the whole nation.

In Table 9.1, we report the population (P) and seats data (H) for the Lower Houses of some countries that hold direct elections of their parliaments. The table also shows how many people are necessary to form a seat (P/H). Barring very large countries where this data is necessarily high, countries with similar populations also have similar values for P/H. It may be noted that the data for Italy is consistent with countries of similar size, notwithstanding the cyclic debates on the idea of there being 'too many' parliament members.

Once the number of seats has been fixed and the district map has been designed it becomes a matter of dividing all seats among the districts according to the populations in each district.

From a mathematical point of view, the problem of dividing the seats at the national level among the various parties to obtain a political representation is equal to the problem of dividing the seats among the districts for a territorial representation.

Table 9.1 P = population, H = number of seats, P/H = seat cost. Population data from the World Bank, 2018 estimates, https://data.worldbank.org/indicator/sp.pop.totl

	P	H	P/H
India (Lok Sabha)	1 352 617 330	552	2 450 390
USA (House of Representatives)	327 167 430	435	752 109
Russia (Duma)	144 478 050	450	321 062
Germany (Bundestag)	82 927 920	709	116 965
France (Assemblée nationale)	66 987 240	577	116 096
UK (House of Commons)	66 488 990	650	102 291
Italy (Camera dei Deputati)	60 431 280	618	97 785
Spain (Congreso de los Diputatos)	46 723 750	350	133 496
Poland (Sejm)	37 978 550	460	82 562
Netherlands (Tweede Kamer)	17 231 020	150	114 873
Sweden (Riksdag)	10 183 170	349	29 178
Austria (Nationalrat)	8 847 040	183	48 344

Instead of considering the present populations, one considers the votes obtained by the various parties.

However, there are features, when seats must be apportioned to parties, that are not observed when seats are apportioned to districts. The requirement of proportionality is central in the apportionment of seats to districts, while it less important in the other case, because we here face another problem, that is, the formation of majorities and minorities, the existence of electoral thresholds and some other possible rules. Moreover, the votes for the parties are often marked with lists of candidates for which a preference can be expressed, and the seats must be allocated not only to the political lists, but also to the candidates according to the obtained preferences.

It seems natural that the allocation of seats to the districts has to be proportional. Indeed, it is almost always so, except in a very special case. The composition of the European Parliament does not follow pure proportionality, but a principle called *degressive proportionality*, that is, the more populous a nation is, the less it is represented, so as to give a greater voice to the smaller states. We will come back to the case of the European Parliament in Sect. 9.5.

Let us thus consider the problem of apportioning seats to districts by adopting the principle of pure proportionality. We denote the population of a certain district with p, the total population with P and the total number of seats with H. If we divide the population by the number of seats (P/H), we obtain the data that indicates how many people are necessary to form a seat. We can call this data *seat cost*.[2] If, instead, we divide the number of seats by the population (H/P), we obtain the reciprocal data that denotes the fraction of a seat 'owned' by each person. We can call this data *representativity*. The higher the representativity, and, consequently, the lower the seat cost, the more the population is represented. By respecting proportionality, the ideal number of seats allocated to a district should be

---

[2]The reader should not be misled by the word 'cost'; it is does not refer to an economic cost, but only numerical data.

$$q = p \cdot \frac{H}{P} = p \cdot R$$

where $R = H/P$ is the representativity. That what is written indeed involves multiplying the fraction of a seat R 'owned' by a person by the number of people p in the district, and the result is the number of seats to which the district is entitled. The number q that is obtained is called the *quota*. Unfortunately, the quota is never integral (it could be, but this is very unlikely), and since the seats must be integer numbers, we have to round the quota.

This may seem to be an insignificant problem, but it is actually a complex one. By necessity, whenever the quota is rounded down, the district is penalized (and the others are, of course, favored because the 'lost' part is given to the others), whereas, when we round the quota up, the district is favored (and the others are penalized). At the very moment when any decision penalizes somebody and favors somebody else, there will always be arguments on how the decision should be made. A 'fair' decision that is above parties does not exist and cannot exist. Various criteria can be invoked, and each one has its legitimacy.

One requirement that seems inescapable consists in demanding that a district cannot receive fewer seats than another district with a smaller population. This requirement is called the *Monotonicity criterion*.

Moreover, it seems correct to demand that the seat number be obtained by simply rounding the quota either down or up. So, if the quota is 13.6, then the seat number should be either 13 or 14, not 12 or 15, let alone 11 or 16. Such a property is called the *Quota satisfaction criterion*.

Another requirement, which seems sensible to be satisfied, takes into account two allocations that are done in different times. Let us assume that the population in one district grows more quickly (in percentage) than the population in another district. When the seats are recomputed a second time, it should not happen that the first district loses seats while the second one gains them. Such a requirement is called the *Population monotonicity criterion*.

The bad news is that there exists no apportionment method that always guarantees the three criteria [17], for the simple reason that there exist cases in which any possible rounding fails to satisfy at least one of the three criteria. By this, we are not saying that it is impossible to satisfy all three requirements in all real cases, but rather that there are some cases in which they are not satisfied, and therefore apportionment methods, which must work in all cases, cannot exist. Such an example can be found in [17]. The methods that are currently used always guarantee two criteria, and, in particular, the Monotonicity criterion, the most important, is always satisfied.

9.2 Largest Remainder Method

An old method that satisfies Monotonicity and Quota satisfaction is the *Largest remainder method*, also known as the *Hamilton method*, or the *Vinton method*, or the *Hare method* or the *Hare-Niemayer method*. This abundance of names all denoting the same method is due to the fact that a number of people, at different times, have all time 'reinvented' the method. The first one to have done so seems to have been anyway the American statesman Alexander Hamilton.

The Largest remainder method is perhaps the simplest way of allocating seats. Initially, the seats are allocated by rounding down the quotas $q = p \cdot H/P$. The difference between the quota and the rounded quota, i.e., the fractional part of the quota, is called the *remainder*. The number of seats that are apportioned in this first phase is almost surely smaller than H (it would be equal only in the very unlikely case of all quotas integral), and therefore some seats remain to be assigned. In order to decide who is entitled to receive these remaining seats, one looks at the districts that have been more penalized by the rounding. Clearly, these are the districts with the largest remainder. So, the remaining seats are allocated one at a time, until completion, to the districts with the largest remainders.

We illustrate the method with a small example. There are three districts with respective populations 5000, 3000 and 1000, among which four seats ($H = 4$) must be allocated. The global population is 9000 people and the quotas are

$$5000 \cdot \frac{4}{9000} = 2.222 \quad 3000 \cdot \frac{4}{9000} = 1.333 \quad 1000 \cdot \frac{4}{9000} = 0.444.$$

Initially, by rounding down, one respectively assigns 2, 1 and 0 seats to the three districts. One seat is left over and is given to the third district, which has the largest remainder (0.444 vs. 0.333 and 0.222). Hence, in total, the seats are 2, 1 and 1. With so few seats, the rounding necessarily produces distortions with respect to the idea of pure proportionality. The seat cost is quite variable among the districts: for the first district, the cost is $5000/2 = 2500$, for the second, $3000/1 = 3000$, and for the third, $1000/1 = 1000$. The people of the third district are 'worth' three times as much as the people of the second. Unfortunately, this is unavoidable when small numbers are rounded (like 0.444), leading to large relative errors.

It is interesting to note that the Largest remainder method is prescribed in the Italian Constitution, both for the House of Representatives (Camera dei Deputati), in art. 56, and for the Senate, in art. 57, where the total number of seats is prescribed. These articles have been amended in 1963 and 2001. In the original formulation (1948), only the seat costs (80,000 and 200,000, respectively) were specified and the rounding was prescribed to the closest integer.

In the Italian Constitution, it is not written as to how the similar task of dividing the seats at the national level among the parties according to the received votes should be carried out. This has always been specified in the electoral laws, which have sometimes adopted the Largest remainder method, as well as divisor methods.

9.3 Paradoxes

The Largest remainder method clearly satisfies the criteria of Quota satisfaction and Monotonicity. It can also be shown that, if we consider 'error', with respect to the ideal fractional quota, the difference, in absolute value, between the quota and the seat actually allocated, then the Largest remainder method produces the allocation with minimum error, no matter how it is measured (either as sums of the errors for all districts or as the maximum error among all districts).

Hence, we might have the impression that the Largest remainder method is the ideal apportionment method, so much so that we cannot ask for more. In fact, there are serious issues.

As a first observation, the Largest remainder method does not take into account the relative error that is made by rounding to the closest integer. For instance, by rounding 2.4 to 2, we make a relative error of $(2.4 - 2)/2.4 = 1/6$, that is, about 16.7%. Contrastingly, rounding 6.4 to 6 (i.e., the same absolute quantity), we make a relative error of $(6.4 - 6)/6.4 = 1/16$, that is, 6.25%, which is much smaller. As we have seen in the small example, the representativity may have a large variation from district to district, whereas, ideally, they should all have the same representativity.

However, even if we are more interested in the absolute error than in the relative error, there is a curious problem that has emerged only on the occasion of apportioning the seats to the House of Representatives of the United States of America (see Chap. 10). It seems sensible that, if the number of seats in the House is increased, no district should lose seats after the recomputation. This should be true for any apportionment method. Let us look again at the previous example of three districts with populations of 5000, 3000 and 1000 for a total of four seats. We have already seen that the quotas and the seats are

$$2.222 \rightarrow 2 \quad 1.333 \rightarrow 1, \quad 0.444 \rightarrow 1.$$

Now suppose that one more seat is added to the House, so that $H = 5$. The computation now gives us the quotas

$$\frac{5}{9000} \cdot 5000 = 2.778 \qquad \frac{5}{9000} \cdot 3000 = 1.667 \qquad \frac{5}{9000} \cdot 1000 = 0.556$$

According to the method, we initially allocate 2, 1 and 0 seats to the three districts. Now, two seats remain and are given to the first and second district, according to the Largest remainder rule. Result: the seats are 3, 2 and 0. The total number of seats has been increased and the third district has lost one seat!

This paradoxical result was discovered after the 1880 Census in the United States, when the State of Rhode Island moved from two seats to one while the total number of seats in the House was increased from 270 to 280. Because of this effect, the US Census Office conducted an evaluation for all values of H from 275 to 350. It turned out that, in going from 299 to 300 seats, the State of Alabama moved from 8 to 7 seats. For this reason, the paradox is universally known as the *Alabama paradox*.

The Alabama paradox is not the only one that troubles the Largest remainder method. Let us consider the following example with three districts with respective populations of 657, 237 and 106 people (for a total of 1000) who have to share a total of 100 seats. With the Largest remainder method, the following quotas and seats are obtained:

$$65.70 \rightarrow 66, \quad 23.70 \rightarrow 24, \quad 10.60 \rightarrow 10.$$

Now imagine that, when the elections are held again after some years, the populations have changed and the new figures are 660, 245 and 105. In total, there has been a slight increase of 10 people, but the third district has lost one inhabitant. Now we obtain the following quotas (rounded to the second decimal figure) and seats:

$$65.35 \rightarrow 65, \quad 24.26 \rightarrow 24, \quad 10.39 \rightarrow 11.$$

Hence, the district whose population is increasing gives one seat to another district whose population is decreasing! This paradox is known as the *Population paradox* and clearly refers to the violation of the Population monotonicity criterion.

But this is not all. Suppose a new district is added to the nation. Then, we add seats to the House in exact proportion to the new population. If we again compute the seats for all districts, including the new one, we would desire that the numbers of seats for the old districts not change and that the new one clearly receive as many seats as the increment. Well, there is no guarantee that this can be granted. This drawback was discovered in 1907 with the entrance of Oklahoma into the United States, and thus this fact is called the *Oklahoma paradox* and also the *New State paradox*.

To see an example, the reader may again consider the previous case with populations of 657, 237 and 106. Suppose a new district is added with 53 people, which amounts to 5 more seats for the House. Computing the seats again, one gets 65, 24, 11 and 5, with a variation between the first and third districts.

The Largest remainder method presents one more defect, which we may refer to as incoherence. After having apportioned the seats, imagine picking up some districts and recomputing the seats for only these districts. As the total number of seats for these districts, we use the sum of the already assigned seats. What we expect from the recomputation is that numbers will not change. In other words, if an apportionment is 'fair', then every part of said apportionment must be fair.

Reconsider the previous example with slightly modified data: three districts with populations of 5000, 3100 and 900 people for four seats in total. The Largest remainder method assigns 2, 1 and 1 seats, respectively. Now pick up the second and third districts and recompute the seats for them by using the data of two seats to be reallocated. The quotas for the second and third districts are

$$\frac{2}{4000} \cdot 3100 = 1.55, \quad \frac{2}{4000} \cdot 900 = 0.45$$

and the methods assigns both seats to the second district (the first written above). So, there is incoherence between the two results.

This 'defect' is similar to the New State paradox, the difference being that districts are now taken out and previously were added.

9.4   Divisor Methods

Due to its many paradoxes, electoral science experts do not consider the Largest remainder method to be a reliable way of apportioning seats. Moreover, this method tries to reduce the absolute deviations from the quotas, while reducing the relative deviations may be judged fair, because this amounts to having representativity values that are as similar as possible among the districts.

To support the idea of considering the relative deviations, we quote the following observation by Balinski and Young ([17] p. 129): "It can be argued that staying within the quota is not really compatible with the idea of proportionality at all, since it allows a much greater variance in the per capita representation of smaller states than it does for larger states". Hence, the Quota satisfaction criterion should not be regarded as an important requirement, simply because, if we measure the fairness of an apportionment by the representativity, i.e., the relative error, it may, in almost a random fashion (this depends on the remainder), either penalize or favor the small districts to a far greater degree than it does the large ones.

The divisor methods may violate the Quota satisfaction criterion, but they do satisfy the Monotonicity and Population monotonicity criteria. In order to introduce the topic, imagine allocating the seats to the districts one at a time by choosing, in turn, that district that is less represented at the moment.

Initially, all districts have zero seats, and so a zero representativity. To increase this value, they will all get one seat in this initial phase. Once all districts have one seat each, their representativity is equal to the reciprocals of the populations. At this point, the least represented district is the largest one, and it will get the next seat. The remaining seats will be allocated, in turn, to the district that exhibits the smallest representativity just before the allocation.

For instance, with three districts A, B and C, respective populations 4000, 5000 and 7000 and $H = 8$ seats, the sequence of seat allocations (after the first three seats) is shown in the following table, in which, in the first column, the seats currently assigned are shown and, in the other columns, the representativity values are multiplied by 2000 (the values are multiplied by 2000 because $R = H/P = 1/2000$ is the representativity at the national level; 8 seats divided by 16000 people, i.e., one seat for every 2000 people), so that the value 1 is the value that corresponds to the national representativity. We denote in boldface the smallest value that determines the choice of the district to which the next seat is apportioned. The last row is the final representativity (values that should be divided by 2000).

H	A	B	C
3	0.500	0.400	**0.285**
4	0.500	**0.400**	0.571
5	**0.500**	0.800	0.571
6	1.000	0.800	**0.571**
7	1.000	**0.800**	0.857
8	1.000	1.200	0.857

So, by taking into account the initial allocation of one seat to every district, we have 2 seats for A and 3 seats for both B and C. It turns out that A is exactly represented (the result of a chance that occurs when dealing with very simple numbers), whereas B is over-represented and C is under-represented.

Clearly, it is impossible that the districts are either all over-represented or all under-represented. Necessarily, some districts will be over-represented and some other districts will be under-represented. By working as we have done, in the very moment that we allocate a seat to a district and this district becomes over-represented, we are sure that no other seat will be allocated to this district, because, otherwise, all districts should be over-represented, and this is impossible. So, we may wonder which district will be so lucky as to end up over-represented. As we have described the procedure, at every seat allocation, the representativity of a district jumps up by the exact reciprocal of its population, advancement that is larger for the less populated districts. Thus, these districts have a larger probability of ending up over-represented.

This way of working, that is, assigning the seats one at a time according to some given rule (the one that we described is just one among many), is viewed in today's literature as an obsolete allocation method. The 'modern' way first computes the quotas for each district as $q = p \cdot R$, exactly as in the Largest remainder method. The two methods differ in the way that they round the quotas. Whereas the Largest remainder method employs two ways of rounding in sequence (first, all rounded down, and then some up), in divisor methods, all quotas are rounded using a common rule (which could be all down, or all up, or all to the closest integer, etc.).

After rounding the quotas, we have to check whether the computed seat values are consistent with the total of H seats. If their sum is indeed H, we have found the apportionment. Otherwise, if the sum is less than H, we have to increase the representativity coefficient R, and if, instead, it is more than H, we have to decrease the representativity coefficient R. In both cases, we compute again the seat numbers and repeat the procedure until a coefficient R is found such that, after rounding, the seat numbers sum up exactly to H. This procedure may look cumbersome, and it certainly is not nearly as immediate as the Largest remainder method, but it can be efficiently implemented on a computer.

The way in which the quotas are rounded is crucial in the definition of a particular divisor method. We have already mentioned some ways of rounding, like rounding down, up, or to the closest integer. But these are not the only conceivable methods. Two more methods determine how to round a quota q that lies between two integers z and $z + 1$. In one method, a threshold value

$$d = \frac{2}{\frac{1}{z} + \frac{1}{z+1}}$$

is computed (d is necessarily between z and $z + 1$) that is used to decide how to round q. If $q < d$, then q is rounded down, and if $q > d$, it is rounded up. What about the case of $q = d$? In this case, we are free to round as we wish, but this event has negligible probability of occurring.[3] For instance, if $2 < q < 3$, we compute

$$d = \frac{2}{\frac{1}{2} + \frac{1}{3}} = \frac{2}{\frac{5}{6}} = \frac{12}{5} = 2.4$$

So, if $q = 2.3 < 2.4$, one rounds q down to 2. If $q = 2.43 > 2.4$, one rounds q up to 3. If $0 < q < 1$, one always obtains $d = 0$, which implies always rounding up a quota smaller than one. In other words, no district can be without seats according to this method.

Another method computes the threshold value

$$d = \sqrt{z(z+1)}$$

In this case, d is also included between z and $z + 1$. If, as in the previous example, we have $2 < q < 3$, we obtain

$$d = \sqrt{z(z+1)} = \sqrt{6} = 2.44949$$

so that $q = 2.43$ is rounded down to 2 according to this method of rounding. Again, a quota smaller than one is always rounded up.

The first three methods that we have shown can also be related to the use of thresholds for rounding. Always rounding down is like having $d = z + 1$, always rounding up corresponds to $d = z$, and rounding to the closest integer corresponds to $d = z + 1/2$. So, we have identified five alternative rounding methods that can be ordered according to increasing threshold values d. The five methods, along with the names by which they are generally known, are:

- $d = z$, *Adams' method* (rounding up);
- $d = 2/(1/z + 1/(z + 1))$, *Dean's method* (rounding according to the harmonic mean);
- $d = \sqrt{z(z+1)}$, *Huntington-Hill or equal proportions method* (rounding according to the geometric mean);
- $d = z + 1/2$, *Webster's method* (rounding according to the arithmetic mean);
- $d = z + 1$, *Jefferson's method* (rounding down).

The question is how large the difference can be by adopting one method instead of another one. There are differences that not very large and yet sufficiently large to cause considerable debates as to which method to adopt. The reader can find the controversial history of the seat apportionment in the U.S. House of Representatives in Chap. 10. Adams' method favors the small districts, whereas at the other extreme, Jefferson's method favors the large districts. To get an idea of the differences caused by these methods, one can look at the example in Table 9.2. The population data (first row) have been randomly generated after fixing $H = 100$ e $P = 10000$. This way, the population data divided by 100 also yields the quota. In the last two rows, we also

---

[3]This freedom is cleverly exploited in the Tie-and-Transfer algorithm for the Biproportional Apportionment problem. We will briefly mention this algorithm on Sect. 12.1.

Table 9.2 Comparison of different divisor apportionment methods

p	111	350	472	598	608	691	807	1092	1097	1142	1363	1669
A	2	4	5	6	6	7	8	11	11	11	13	16
D	1	4	5	6	6	7	8	11	11	11	13	17
H	1	4	5	6	6	7	8	11	11	11	13	17
W	1	3	5	6	6	7	8	11	11	11	14	17
J	1	3	4	6	6	7	8	11	11	12	14	17
R	1	3	5	6	6	7	8	11	11	11	14	17
I	0	3	4	6	6	7	8	11	11	12	14	18

report the apportionments for the Largest remainder method and for the Imperiali method, which we will illustrate next (A = Adams, D = Dean, H = Huntington-Hill, W = Webster, J = Jefferson, R = Largest remainder, I = Imperiali).

The name 'Divisor methods' comes from the fact that these methods were first defined by assigning a particular sequence of integer numbers (like 1, 2, 3, 4, ..., for instance), called *divisors*, and then building a table in which, for each district, the population data is divided for each one of the divisors. For the previous example, with three districts, respective populations of 4000, 5000, 7000 and $H = 8$ seats, by adopting the divisors 1, 2, 3, 4 and 5, we have the table

divisors	1	2	3	4	5
A	**4000**	**2000**	1333	1000	800
B	**5000**	**2500**	1666	1125	1000
C	**7000**	**3500**	**2333**	**1750**	1400

where, in boldface, the eight (like H) largest numbers are put into evidence. These determine the district to which the seats are assigned. So, two seats go to A and B and four seats to C.

The reader will certainly notice that this manner of assigning seats is very similar to the one that we have previously shown of assigning seats one at a time to the district with the currently smallest representativity. In this table, we consider the inverse ratios, i.e., the costs of the seats, but taking into account the largest cost is the same as taking into account the smallest degree of representativity. In the previous presentation, there was an initial allocation of one seat to each district, which we don't see in this case. In particular, with respect to the previous case, we evaluate the representativity of a district as if the district were to have one more seat than those actually assigned.

We may also add the number zero to the divisors 1, 2, 3, ... This would imply a division by zero, and thus infinite values in the table. If we admit these values and consider them, we have the exact apportionment shown earlier.

Is there any relation between this way of assigning seats and the procedure that we have previously outlined? It can be shown (although it would take too long to do so here) that they are, indeed, the same thing. In particular, the choice of 1, 2, 3 etc., as divisors, proposed by D'Hondt (and thus called *D'Hondt's method*) corresponds to Jefferson's method, and the choice of 0, 1, 2, . . ., corresponds to Adams' method.

These divisors are not the only ones that have been proposed. For instance, the sequence of odd numbers 1, 3, 5, 7, . . ., has also been considered. This method is known as the *Sainte Laguë* method. By applying these divisors to the example, we obtain

divisors	1	3	5	7
A	4000	1333	800	444
B	5000	1666	1000	555
C	7000	2333	1400	777

So, two seats go to A and three to B and C. It can be shown that these divisors correspond to Webster's method. If we change the first divisor from 1 to 1.4, we have the so-called *modified Sainte Laguë method*. This method corresponds to having a threshold value $d = k + 1/2$ for all values of k except $k = 0$, for which we define $d = 0.7$. In this way, small districts are penalized to a greater degree.

The *Belgian method*, also known as the *Imperiali method*, has divisors 1, 1.5, 2, 2.5,. . ., and the *Nohlen method* has divisors 2, 3, 4, 5,. . .. In the second method, the divisors are double those in the first one. Hence, they produce the same result, because it is just a matter of rescaling all of the numbers. It can be shown that these two methods involve considering a threshold value $d = z + 2$ for rounding. It is quite inappropriate to speak of 'rounding' in this case. It would be like saying that a quota of 3.3 is rounded to 2 or a quota of 5.6 is rounded to 4 (although 0.4 would be rounded to 0 in any case).

If Jefferson's method favors large districts, the Imperiali method favors them even more (see Table 9.2). We have to say that this method is mainly used in assigning seats to lists, rather than to districts, so as to penalize least voted-for lists (as an alternative to electoral thresholds).

Going back to the idea of assigning seats one at a time to the district with the smallest degree of representativity, the Imperiali method involves evaluating districts, as if we had already assigned two seats each, but without having apportioned these two seats. It is clear that very small districts for which two seats would be sufficient, are already 'represented' by two virtual seats without having any real seats. This is the result of the Imperiali method applied to the previous example

For this example, we obtain the same apportionment as in Jefferson's method (or D'Hondt's) with a clear imbalance in favor of the largest district. Note that, while with Jefferson's method, the last seat is assigned to C, with Imperiali's is assigned to A. So, if there are seven seats instead of eight, district A would loose the seat with

divisors	2	3	4	5
A	2000	1333	1000	800
B	2500	1666	1250	1000
C	3500	2333	1750	1400

Imperiali and C would with Jefferson. The Imperiali method has another 'oddity' that is not present in any other method. If, by chance, the quotas are integral, there would be no need for rounding, and all methods would give the same apportionment (equal to the quotas). For instance, with three populations of 3000, 5000 and 7000 that have 15 seats to share, the exact proportionality gives us integral results, i.e., 3, 5 and 7 seats. These numbers would be obtained with all of the methods, except Imperiali, which gives us 2, 5 and 8 seats (as the reader can easily compute).

9.5 Seats of the European Parliament

We have previously mentioned that the seats in the European Parliament are not apportioned according to a proportionality principle, but rather to a degressive proportionality principle, that is, larger countries are less represented. Among the European countries, there are large differences in populations and so it seems fair to avoid the possibility that large countries might prevail easily over small countries. The basis of this principle is obviously the idea that territorial representativity is worth much more in the European Parliament than in a national parliament.

The requirements for the seat apportionment are as follows:

1. no country should receive fewer seats than a smaller country;
2. no country should receive fewer seats than a given lower threshold;
3. no country should receive more seats than a given upper threshold;
4. seats must obey the degressive proportionality principle.

Degressive proportionality means that the population/seats ratio, i.e., the seat cost, must increase with the population (instead of being constant, or almost constant, as pure proportionality requires). Apparently, this criterion explicitly favors small countries. The data from the European Parliament indicate 751 seats for the 2014 election and 705 seats for the 2019 election. The reduction is essentially due to Brexit. The 73 seats formerly apportioned to the UK have been partly (46 seats) dropped and kept for future additions into the Union and partly (27 seats) redistributed among the actual countries. The lower threshold is 6 seats and the upper threshold is 96 seats.

The four listed rules raise some problems as to how to compute the seats. A committee designated to solve the question had proposed a particular divisor method [46]. The proposal called for an initial allocation of six seats to each country and then the use of a divisor method for the remaining seats, with the allocation possibly being blocked in case the upper limit is reached.

The imposed constraints considerably restrict the range of possible solutions, and there may be problems for countries with comparable populations. In this respect, an example is shown in [46] in which, by simply extracting five states with almost equal populations, it can be seen that no apportionment exists. Clearly, politicians have imposed rules that require a certain degree of mathematical competence for their satisfaction. In any case, the problem is not insurmountable and can be faced in several alternative ways. A number of suggestions as to how face the problem are featured in [57]. In particular, new quotas are defined in [87], called 'projective', that respect the requirements.

Oddly enough, it does not seem that mathematicians were present in the final stage of seat computation, given that the degressive proportionality requirement is satisfied neither for the 2014 election nor for the 2019 election. The graphs in Figs. 9.1 and 9.2 show the seat cost as a function of the population for the 2014 and 2019 parliaments, respectively (note that the UK is missing in the second graph). The dots represent the pair 'population–seat cost' for each country in the European Union. We also show a curve that interpolates the dots.

According to the degressive proportionality principle, the points should be placed at increasing heights moving to the right. As can be seen from the graphs, the requirement of degressive proportionality is not respected in a few cases.

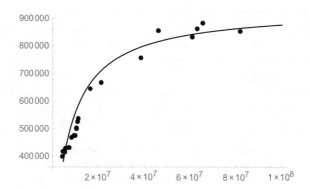

Fig. 9.1   Seat costs versus populations for the European Parliament, 2014

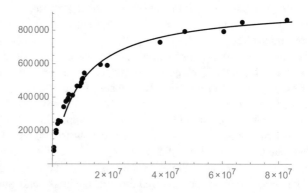

Fig. 9.2   Seat costs versus populations for the European Parliament, 2019

Chapter 10
The United States Congress

The quite controversial history of the various decisions that have been made over the course of 200 years to apportion the House of Representatives of the United States of America, is a very good indication of all of the difficulties inherent in the issue. Article 1, Section 2, of the U.S. Constitution states, in its original version:

Representatives and direct Taxes shall be apportioned among the several States which may be included within this Union, according to their respective Numbers, which shall be determined by adding to the whole Number of free Persons, including those bound to Service for a Term of Years, and excluding Indians not taxed, three fifths of all other Persons. The actual Enumeration shall be made within three Years after the first Meeting of the Congress of the United States, and within every subsequent Term of ten Years in such Manner as they shall by Law direct. The Number of Representatives shall not exceed one for every thirty Thousand, but each state shall have at Least one Representative; and until such enumerations shall be made...

The article goes on by directly listing the number of representatives for each one of the 13 original states, for a total of 65 representatives. This number was in effect until the first census, which was held in 1790, three years after the approval of the Constitution on September 17, 1787. According to the article, in order to determine the seats, one does not count the inhabitants directly. One has to distinguish between the 'free Persons' who count as one, the Native Americans ('Indians') who count as zero because they don't pay taxes, and the slaves ('all other Persons') who count as only three fifths. The electoral numbers that are so obtained ('their respective Numbers') are then used to determine the seat numbers. This article was modified in 1868 after the abolition of slavery by the 14th Amendment.

In the first census (1790), the count was of 3,199,357 free persons and 694,280 slaves (the Native Americans are not even counted, although it probably would have been very difficult to take their census anyway) [99, 101]. Therefore, the electoral number to consider was $3,199,357 + 3/5 \cdot 694,280 = 3,615,925$. Given the requirement that there be, at most, one representative for every 30,000 electoral numbers, the maximum number of seats was 120 (which yields a ratio of 30,132

© Springer Nature Switzerland AG 2020
P. Serafini, *Mathematics to the Rescue of Democracy*,
https://doi.org/10.1007/978-3-030-38368-8_10

numbers/seats), and this value was chosen by the Congress in 1791, thus respecting the constitutional provision by a small margin.

Moreover, the Congress approved apportioning the seats according to the Largest remainder method, following the proposal by Alexander Hamilton. The proposal by Thomas Jefferson, which called for the method that bears his name, was rejected. However, President George Washington exercised his veto power, the first time that power had been used in U.S. history. The Congress had to approve a new law that provided for 105 seats to be apportioned according to the Jefferson's method.

The principal reason for the veto lay in the number of seats that, by having almost exactly 30,000 electoral numbers per seat, necessarily violated the constitutional requirement in some states (specifically eight) because, as we have seen, no method can avoid having some states under-represented and some others over-represented. However, this was according to the interpretation of Jefferson[1] and Edmund Randolph, whereas, according to Hamilton and Henry Knox, the constitutional requirement had to be considered for the nation as a whole, and not just state by state. According to Knox, it was not clear "whether the numbers of representatives shall be apportioned on the aggregate number of all the people of the United States, or on the aggregate numbers of the people of each state".

Less evident was the veto for unconstitutionality against Hamilton's method, which, according to Jefferson, was even "repugnant to the spirit of the constitution", since, according to his interpretation, the constitutional provisions implied the use of divisors (see the previous footnote). Hamilton disagreed. This was just one among the many disputes between the two politicians.

Eventually, Washington took sides with Jefferson and vetoed the bill.[2] We may wonder why Washington made such a decision, beyond the unconvincing official reasons. We should dismiss the notion that he was aware of the Alabama paradox (although, even if Alabama was not yet included in the Union, Washington could have had knowledge that there was a problem). We may conjecture that both Washington and Jefferson, coming from the populous Virginia, preferred this method that favored their state (indeed, Virginia received 19 seats by Jefferson's method and would have received only 18 by Hamilton's method). In any case, there is no evidence in favor of this conjecture.

As already explained, the representativity had to be no larger than $1/30,000$. Since then, the representativity has decreased considerably over the years, because the total population has increased much more quickly than the number of seats, and so the constitutional provision is amply satisfied today. The number of seats has been stable at the value of 435 since 1910. This value implied a cost of 210,328 citizens per seat in 1910, whereas, in 2010, the cost is 710,767 citizens per seat

---

[1] "I answer, then, that taxes must be divided exactly, and representatives as nearly as the nearest ratio will admit; and the fractions must be neglected, because the Constitution calls absolutely that there be an apportionment or common ratio, and if any fractions result from the operation, it has left them unprovided for" [52].

[2] "[T]here is no one proportion or divisor which, applied to the respective numbers of the States, will yield the number and allotment of Representatives proposed by the bill" [103].

($= 309, 183, 463/435$, the population data does not include the inhabitants of the District of Columbia, which does not have representatives in the Congress) [100].

Since it had become quite manifest that Jefferson's method favored the largest states, various proposals for change were presented. In 1822, William Lowndes proposed a method of his own invention, a mix between divisors and Largest remainders, but the proposal was rejected. In 1832, John Quincy Adams proposed the method that bears his name, and Daniel Webster later proposed his own method. Both proposals were rejected, at least in part for the obvious reason that the House was mainly represented by the largest states. Jefferson's method was maintained, with an increase in the number of seats to 240.

However, in 1842, Jefferson's method was eventually abandoned in favor of Webster's with a decrease in the number of seats to 223, although, by 1852, Webster's method had already, in turn, been abandoned in favor of Hamilton's method. The proponent was Samuel Vinton, so that Hamilton's method is also known as Vinton's method. The seat number was changed to 234, because, with that number, both Hamilton's method and Webster's method gave the same result.

In 1872, the seat number was increased to 283 so as to get the same apportionment with both methods, but 9 other seats were also added and the resulting apportionment, decided after a fierce political battle, did not coincide with any previous method. The issue became critical in 1876 with the Presidential election. The electoral votes for the Presidential election are based on the House seats. Hayes won with 185 votes against the 184 gained by his rival Tilden. If the apportionment had been the one computed by the previous methods, Hayes would not have won.

In 1880, the Alabama paradox emerged and in 1882, in order to avoid the malfunction of Hamilton's method, the seats were increased to 325, so as to have the same apportionment with both Hamilton's and Webster's methods. In 1890, the seats were increased to 356 with the provision that it always be kept in mind that the two methods had to agree. It is noteworthy that the decision about the number of seats was based more on an equity criterion (if two different methods give the same results, this should be considered a sign of equity) than on a 'political' convenience.

In 1901, the Census Office carried out a preventive computation for all values of H from 350 to 400. The Alabama paradox emerged again, this time involving Maine and Colorado. In particular, for the value $H = 357$, Colorado would have received two seats, whereas, for all other values, it would have received three seats. Quite oddly, the president of the committee for the apportionment proposed the precise value of 357, causing a violent uproar in the Congress, which, of course, rejected the proposal, abolished Hamilton's method and adopted Webster's method with 386 seats.

Even though Webster's method was adopted, a parallel computation was also carried out according to Hamilton's method so as to compare the results, and so, in 1907, the New State paradox was discovered, when Oklahoma was inducted into the Union. Because of this fact, Hamilton's method was definitively shelved and, in 1910, Webster's method was reconfirmed by fixing, for the last time, the number of seats to 433, plus two more seats for allocation to Arizona and New Mexico at the time of their induction into the Union.

In the years after 1920, a study was conducted in the interest of modifying Webster's method for the sake of better equity. Joseph Hill, from the Census Office, proposed a method that was later perfected by a school mate of his, the mathematician Edward Huntington [51]. This method was called the Equal Proportions method by Huntington, but is unsurprisingly also known as the Huntington-Hill method.

Eventually, in 1929, the problem was passed to the National Academy of Sciences, which entrusted a group of mathematicians, formed by G.A. Bliss, E.W. Brown, L.P. Eisenhart and R. Pearl, with the task of studying the problem. The proposal was to adopt the Huntington-Hill method. The same recommendation was made in 1949 by another authoritative group of mathematicians composed of none other than Harold Marston Morse, John Von Neumann and Luther Eisenhart himself.

Finally, in 1941, the Congress adopted the Huntington-Hill method and fixed the number of seats to 435. The method is still in effect.

In 1991 and 1992, Montana and Massachusetts initiated separate lawsuits in district courts challenging, for the first time in U.S. history, the constitutionality of the current method. The question was brought all the way up to the Supreme Court. Montana even challenged the constitutionality of the automatic apportionment law, with the motivation that it deprived the Congress of the right to vote on the matter!

The Supreme Court rejected the Montana claim (the Massachusetts claim had already been dismissed in the district court). Justice Stevens found that an automatic use of an otherwise constitutional apportionment method is a sensible procedure that removed apportionment from political controversy. He found nothing in the Constitution that prevented the adoption of an automatic procedure, and concluded his opinion on March 31,1992, putting an end to a 200-year-long debate, by writing: "The decision to adopt the method of equal proportions was made by Congress after decades of experience, experimentation, and debate about the substance of the constitutional requirement. Independent scholars supported both the basic decision to adopt a regular procedure to be followed after each census, and the particular decision to use the method of equal proportions. For a half century the results of that method have been accepted by the States and the Nation. That history supports our conclusion that Congress had ample power to enact the statutory procedure in 1941 and to apply the method of equal proportions after the 1990 census."

Chapter 11
Legislative Political Representation

11.1 Foreword

There are two antithetical methods for determining how many seats a party (or an electoral list) should receive in the parliament. In one method, one or more seats are available in every district, for instance, with uninominal or plurinominal constituencies. In every district, candidates are elected and, according to the party membership of the elected candidates, the number of seats to which each party is entitled is consequently determined. The British system works this way, with uninominal constituencies, and the method is amply employed, with uni- or plurinominal constituencies, in all Anglo-Saxon countries and many of those with a historical British influence.

We have already remarked (see Sect. 9.1) that uninominal methods can be very distortive with respect to the popular vote. The same problem can also be observed with plurinominal constituencies when the final seats of the parties depend on the elected candidates.

The other method, which is more adherent to the popular vote, determines the votes to be allocated to the parties on a national basis, independently of the preferences (either expressed or not). The way in which to apportion the seats at the national level to the various lists can be done, in principle, with any one of the methods that we have presented in Chap. 9. However, in a parliamentary democracy, in which the parliament is also entitled to express a vote of confidence to the government, several expedients may come into play for the purpose of avoiding fragmentation into too many parties or advantage for those parties (or the coalition of parties) that get more votes. Among these devices, we recall electoral thresholds below which seats are not assigned and majority bonuses that artificially increase the votes to transform a simple majority into an absolute one.

Here, we do not deal with these techniques. The aim of this volume is to provide a theoretical basis for the difficulties that are intrinsic to the very concept of collective choice and that persist no matter which electoral system one has in mind. For this reason, we are not presenting the various electoral systems that have been employed,

P. Serafini, *Mathematics to the Rescue of Democracy*,
https://doi.org/10.1007/978-3-030-38368-8_11

for instance, in Italy (in too many, the electorate must have familiarity with a voting method and the consequent outcome).

Summing up, in the first method, candidates are voted on and, from the elected candidates, a political representation in parliament is formed, and in the second method, parties are voted on and, from the seats allotted to the parties, one determines the elected candidates. In addition, mixed systems can be employed, in which some constituencies may follow one method and others the other method.

That which we are going to present applies to both the first method and the second method. The number of available seats is that of the whole district in the first case and only that of the seats apportioned to a party in a district in the second case. However, we will make explicit reference to the second case, which is closer to the Italian system.

11.2   Choice of the Candidates

We assume that both the number of seats for each list at the national level and also the apportionment of seats to each list in every district have been already decided. How to carry out this second apportionment, an absolutely non-trivial problem, will be the subject of Chap. 12. Now, we face the following problem: to which candidates should we assign the seats gained by a certain list in a certain district?

There is a very simple method for answering this question and it is presently in effect in Italy: a ranking is formed a priori and the seats are assigned in sequence following the ranking. The voters, by their votes, have only decided how many seats are apportioned to the list, but the names have already been decided by the parties. If we add to this fact the almost ubiquitous faculty (at least for the most influential members of a party) to be candidates in several districts, it appears evident that the choice of candidates becomes an exclusive trait of the parties and may also turn into a bargaining chip.

It was argued, and not incorrectly, that the preference mechanism allowed for 'signing' the electoral ballot, and thus for controlling the vote. To some extent, this is true (see also our footnote on Sect. 8.2). If we have only three candidates A, B and C, there are 16 different ways of indicating the preferences: none, A, B, C, AB, BA, AC, CA, BC, CB, ABC, ACB, BAC, BCA, CAB and CBA. The number of different possibilities grows very quickly with the number of candidates.[1] With six candidates, the number is already 1957, sufficiently high to univocally 'label' all voters of a particular polling station. But with four preferences (as was the case in

---

[1]To people who find amusement in mathematics, we may say that this number grows more quickly than the factorial and is given by the following recursive formula: $f_n = n\, f_{n-1} + 1$, with $f_0 = 1$. Asymptotically, it is equal to the factorial times the number e.

Italy until 1992), the number is 65, sufficiently low to avoid manipulations,[2] and we could get back to this number to grant freedom of choice to the voters.

So, let us assume that the voters have expressed certain preferences in regard to the candidates and, from these preferences, we have to choose some people. The problem is not very different from that of determining the winner of an election. The difference is that, instead of one winner, we have more than one winner. Hence, we have to reconsider the ideas that we have already presented and extend them to the case of a multiple choice. Of course, we expect the same conceptual difficulties that are present for a single choice.

Right away we present a paradigmatic example that is useful for highlighting the problematic aspects of the choice. There are 4 candidates, 14 voters and 2 seats to allot. As usual, we imagine that voters have precise rankings in mind. The situation is as follows:

$$5 \text{ electors:} \quad A \rightarrow B \rightarrow C \rightarrow D$$
$$5 \text{ electors:} \quad D \rightarrow C \rightarrow B \rightarrow A$$
$$2 \text{ electors:} \quad B \rightarrow C \rightarrow A \rightarrow D$$
$$2 \text{ electors:} \quad C \rightarrow B \rightarrow D \rightarrow A$$

which yields the Condorcet table

	A	B	C	D
A		5	5	7
B	9		7	9
C	9	7		9
D	7	5	5	

There is no Condorcet winner. However, since we have to express two winners, we see that B and C are even and prevail over both A and D. Hence, the seats are given to B and C. If we take into account the Borda score, we get the same conclusion: 25 points for B and C and 17 for A and D.

The example might correspond to two extreme candidates, although of opposite leanings, A and D and two moderate candidates B and C, with B being closer to A and C closer to D. The five voters who prefer A abhor D and the other five voters feel the exact opposite way. Then, there are four voters who prefer the moderate candidates, two with political leanings like B and two like C.

---

[2]However, if we leave freedom to indicate a candidate in three alternative ways (no first name, only initials, full first name) then the number grows vertiginously as the recursion $f_n = 3n\, f_{n-1} + 1$, $f_0 = 1$, which for $n = 4$ already yields 2713.

Let us assume now that each voter is allowed to indicate only one preference. So, the votes that are counted are: 5 each for A and D and 2 each for B and C. The simplest way to proceed, once the preferences have been counted, is to allot the seats following the preference ranking. This method is called *Single Non-Transferable Vote, SNTV*. The reason for the term 'non-transferable' will be clear in a moment. This is exactly the mechanism of simple majority applied to more people. All of the objections that were raised against the simple majority principle also remain valid with multiple choice.

As already remarked, this method is also used to elect all candidates of a constituency directly, possibly related to different parties. It goes without saying that the problem of vote-splitting is crucial when working in this way and voting inevitably becomes strategic. If, instead, the method is used within the seats assigned to a party, the vote-splitting problem is less relevant, but the outcome can still be problematic. In the example, A and D would be chosen with 5 votes each, while B and C, Condorcet winners, would be discarded!

If we want to evaluate the preferences in a more reasoned way, we may adopt the logic of absolute majority by adapting it to the case of more winners. The logic of the absolute majority for a single winner can also be reformulated by saying that it must be impossible for all non-chosen candidates to overcome the winner, even if they form a coalition, and, to this purpose, the most voted-for candidate must earn more than half of the votes.

Suppose now that there are two seats to be assigned. The non-chosen are therefore all of the candidates except two and, of course, they should be the candidates from the third to the last position according to the preference ranking. If these candidates, by summing their votes, do not earn more than one third of the votes, they cannot overcome the first candidate. Indeed, if they have, at most, one third, the first two together have at least two thirds and the best candidate has at least one third. So, the first winner can be chosen according to the majority principle, and if the second also has more than one third of the votes, they can both be chosen. If, instead, the second does not pass the threshold of one third, the decision could still be made to elect the second, but this decision would no longer be based on the majority principle.

So, according to this logic, if there is one seat to be assigned, the threshold to pass is one half, and if there are two seats, it is one third. It is not difficult to become convinced that, if there are three seats, the threshold must be one fourth, and if there are four, it must be one fifth, and so on. This type of threshold is denoted as the *Droop quota*, and, since we have to consider integers (the numbers of votes are integral), it is computed as the upper rounding of the number of votes divided by the number of seats increased by one.[3]

In the example, both A and D have more than one third of the votes (they have the exact Droop quota) and so, following this logic, they are entitled the two seats without doubt. Indeed, they have 5 votes each and B and C obtain only four votes by summing their votes. Hence, even if we try to extend the concept of absolute majority

---

[3]If, by chance, the division were to yield an integral number, one must add one.

to the case of more seats, we obtain a result that stands in contrast to Condorcet's criterion. This did not happen with a single seat.

In case there are not enough candidates beyond the threshold to assign all available seats and we still want to invoke a majority principle to elect all of them, by necessity, we have to add more information to the vote. The variant known as *Single transferable vote*, which indeed requires more information, is very popular mainly in nations of British heritage: each voter indicates not only their first choice, but also their second, their third and so on until a fixed number is reached (not too high, because it is doubtful that electors can go much further with their choices). In any case, a voter is not compelled to indicate their subsequent choices.

The method follows the same logic of the 'majority threshold' that we have previously presented. The new element is that the elected candidates may, so to speak, 'share' their preferences according to the indications of the voters. Initially, all candidates who pass the quota are elected. The additional idea is that the votes beyond the threshold are not really necessary to get the seat, and so they can be transferred to some other candidate (this explains the name of the method). Of course, the choice of the candidate to whom some votes have to be transferred cannot be arbitrary. The rule is that the votes are transferred to the second choices. This is the same mechanism as in Instant run-off voting (presented in Sect. 6.3) amplified to more candidates.

For instance, if candidate A has received 30 votes and the quota is 20, those 10 extra votes must be given to all of those candidates who are second choices when A is the first choice. Clearly, it is not possible to identify the 10 extra votes among the 30 votes received by A. Hence, the 10 votes must be 'spread' over all 30 for a value of $10/30$ votes. These $10/30$ votes go to every second choice (when A is the first choice). If, for instance, B is the second choice in 8 votes, C in 15 votes and D in 7 votes, the votes for B, C and D are respectively increased by $8 \cdot 10/30 = 8/3$, $15 \cdot 10/30 = 5$ and $7 \cdot 10/30 = 7/3$. The sum of these added votes is exactly equal to the 10 votes 'taken' from A and the total sum is not altered.

After the votes have been transferred, it is checked as to whether there are any candidates beyond the quota. If there are none, the least voted-for candidate is eliminated and his/her votes are transferred to the second choices. This method goes on until all seats are assigned using the various choices (second, third, etc,) depending on the situation. If all of the choices of a vote have been used up, that vote is discarded from the count.

Apparently, the method is quite laborious. Essentially, all of the votes that are transferred are those that are beyond the quota for the elected candidates and all of the votes of the eliminated candidates. The problem does not so much lie in the complex mechanism, which does not make the final outcome immediately visible, but rather in its logic. Though invoking the absolute majority principle, the method may produce an outcome in deep contrast with the outcome that we would have obtained if all elector rankings had been available. See the previous example, in which the transfer of votes was not even necessary. What we have shown here is the same as what we demonstrated when dealing with the difficulties of Simple majority.

To give further evidence of the problems inherent in this method, we consider again the previous example on Sect. 11.2, and try to evaluate the candidates by using the Majority judgement. Of course, we do not have the judgements, but we may adapt the ranking by inventing the judgements of First, Second, Third and Fourth, which we might also interpret as Good, Acceptable, Poor and to Reject, according to the presentation of this example.

	First	Second	Third	Fourth
A	5	0	2	7
B	2	7	5	0
C	2	7	5	0
D	5	0	2	7

The majority grades are Fourth+ (to Reject+) for both A and D and Second− (Acceptable−) for both B and C. Therefore, the seats are given to B and C, according to the Majority judgement, as they would also be with Condorcet and Borda.

Finally, another simple method for the choice of candidates can be based on Approval voting. The names in the electoral ballot are not written in order of preference, but they are all approved at equal merit. All preferences are counted and the candidates are chosen according to the ranking. This was the way that preferences were evaluated in Italy, at the time this method was in use. The problem of vote-splitting is mitigated, but all of the problems connected with Approval voting remain. If, in the example, we suppose that the extreme voters indicate only one preference, either A or D, just because they are extreme, and, contrastingly, the moderate voters indicate the two central choices B and C, then we would count 5 votes each for A and D and 4 each for B and C. A and D would win again.

There seems to be no way out of this situation if we don't add information. Even if the additional information is related to the second, third and subsequent choices, the outcome may not be satisfactory, as in the example in which A and D are chosen instead of B and C. Since the indication of several choices requires some effort on the part of the voter, we might as well ask for some extra effort to express the judgements required by Majority judgement. We believe that the solution to the problem of choosing the candidates could be Majority judgement, which can also be used to formulate a ranking, as demonstrated at length in Chap. 8.

We should mention that the choice of certain persons out of a group of candidates, is also present in contexts that are different from the typically proportional one that we find in political elections. The choice of committees of experts belongs within this type of consideration, but follows somewhat different logic. The interested reader can find a detailed analysis in [35].

Chapter 12
Biproportional Representation

12.1 Biproportional Apportionments

We have seen in the previous chapters that parliament seats must be divided among districts following a proportionality criterion with respect to the populations and, at the same time, they must be divided among the political lists according to the votes received by the lists at the national level. In this chapter, we deal with the problem in which the seats apportioned to the districts have been determined before the election, the seats apportioned to the lists have been computed after the election and we still have to apportion the seats to every list in every district. It is obvious that this apportionment must respect the total number of seats, both for every district and for every list.

In these electoral systems, the vote expressed to a list is prominent and the choice of representatives is subsequent to the seats apportioned to the lists. In some systems, the seats assigned to the lists are, contrastingly, consequence of the seats won by the candidates (like in the uninominal British system).

A fundamental requirement is that the seats be as proportional as possible to the expressed votes. This problem is indeed called *Biproportional apportionment*. Luckily, it is a problem whose various formulations can be solved in an algorithmically efficient way. However, neither is it a problem so simple that it can be approached without adequate mathematical knowledge and tools. Indeed, in some cases, anomalies have been observed in the procedures prescribed by the electoral laws that have been caused by conceptual mistakes. The most striking case is Italy, where noteworthy anomalies took place in the political elections of 1996, 2006, 2008 and 2013. We will address this case in detail at the end of the chapter.

What needs to be computed is a table of integer numbers, with the rows corresponding to the districts and the columns to the lists. Each number represents the number of seats that have to be assigned to a certain list in a certain district. This table must satisfy the following requirements:

– the sum of the number in every column (i.e., for every list) must be equal to the number of seats already assigned to that list;

© Springer Nature Switzerland AG 2020
P. Serafini, *Mathematics to the Rescue of Democracy*,
https://doi.org/10.1007/978-3-030-38368-8_12

- the sum of the number in every row (i.e., for every district) must be equal to the number of seats already assigned to that district;
- if a list has received zero votes in a district, it must receive zero seats in that district;
- the seats should be "as proportional as possible" to the votes.

Whereas the first three requirements are not ambiguous, the last one can be expressed in several alternative ways, each one of which presents both positive and negative aspects. The problem does not only lie in the fact that we want integer numbers, and we have seen how complex the problem of rounding is. Even if the seats could be fractional, it is not at all obvious what 'as proportional as possible' means, due to the double proportionality required with respect to both the districts and the lists.

As in the simple case of territorial representation, we use the term *quotas* to refer to fractional numbers that represent the ideal seat apportionment without the integrality requirement. Whereas, in the proportional apportionment problem, the definition of quotas does not normally constitute a problem (beside the case of the European Parliament), now the question is different.

The first idea could be to define, as has been done in the simple proportionality case, a quota q for a certain list in a certain district by the formula

$$q = v \, \frac{H}{V}$$

where v is the votes obtained by the list in that district, V is the total votes (for all lists in all districts) and H is the number of seats in the House. The ratio H/V defines the representativity at the national level, i.e., what fraction of a seat is owned by each voter, and this data must be multiplied by all votes obtained by a list in a district.

With these quotas, one can build a table of fractional numbers with rows corresponding to districts and columns to lists. This table represents the exact proportionality. This way, the third and fourth requirements are satisfied. However, it is almost sure that the first and second requirements (sums over the rows and over the columns) are not satisfied.

Whereas the first requirement could not be verified by a small margin (after all the votes for each list are computed starting from the votes that are obtained in the districts), the second requirement may give rise to a large discrepancy between the number of seats apportioned to the district and the quota sum. The reason for this is that the seats are apportioned to the districts before the elections on the basis of the actual populations, whereas the quotas are computed according to the number of voters. The two sets of data differ not only because people with the right to vote are always one part of the population, but also because participation in the vote may vary considerably from district to district.

The quotas that are in use in various nations, including, for instance, Italy and Belgium, are the so-called *regional quotas* that are independently computed in each district. Hence, if HD is the seats apportioned to a district and VD is the total votes expressed in that district, the quota is computed by the formula

$$q = v \, \frac{HD}{VD}$$

In this way, the proportionality is maintained only within a district, and so the global proportionality is lost, but the second requirement is clearly satisfied.

Is it possible to define quotas that also satisfy the first requirement, but are less proportional? The answer is affirmative. Such quotas are called *fair share* and, in the literature, they are considered the ideal quotas (i.e., 'fractional' seats).

Their computation is not very difficult, but it would be too complex to be presented here. We limit ourselves to sketching as to how it works. First, we operate on the rows. Each row is multiplied by a coefficient (one for each row) so that the sums in the rows correspond to the required number of seats. Then, the procedure is repeated on the columns, with coefficients for each column. After this operation, the sums on the rows have been spoiled, and so they have to be adjusted again. Now, the sums on the columns are spoiled, and they too have to be adjusted. The procedure goes on, alternating rows and columns. Clearly, the procedure might never stop. However, if some mild conditions are satisfied, the errors in regard to the sums become smaller and smaller, and, eventually, there is convergence toward a table of values that represent the fair share quotas. This procedure can be done very quickly on a computer.

We want to point out an important issue with respect to the definition of a quota. If quotas must respect both sums, any small vote change in some district inevitably induces changes for all districts and lists. If, contrastingly, we prefer that there be some autonomy among districts, in the sense that votes in one district do not influence the seat allocation in some other district, then regional quotas seem more appropriate than fair share quotas.

However, fair share quotas enjoy important mathematical properties that affect the quality of the apportionment with respect to the quotas. For instance, staying within the quotas is also an important property for the biproportional apportionment (once the chosen quotas are acknowledged as 'ideal'), and fair share quotas guarantee the existence of an apportionment within the quotas. This guarantee is important if we look for an apportionment whose seats are simply obtained by rounding the quotas, either up or down. By using regional quotas, such a guarantee is not present, and there may be situations in which there is no apportionment within the quotas.

For instance, this happens with the data in Table 12.1 (the example from [89, 90]). We report the votes in five districts ($D1$, $D2$, $D3$, $D4$ and $D5$) for six lists ($L1$, $L2$, $L3$, $L4$, $L5$ and $L6$), the seats allocated to the district (last columns) and the seats allocated to the lists (last row). These last values can be obtained from the row above either by the Largest remainder method or by Jefferson's or Webster's methods. The example is built so that quotas can be directly read from the votes by dividing them by 10,000. The quotas all lie within the interval 0–1. For instance, the value 9,920 yields a quota of 0.992, whereas the values 10 yield a quota of 0.001. Moreover, the seats apportioned to the districts are exactly proportional to the votes. This gives us the same abstentionism rate among all districts. In other words, the lack of a solution is not due to anomalous data.

Table 12.1   A case of necessary quota violation

	L1	L2	L3	L4	L5	L6	Totals	Seats
D1	9,920	8,700	1,700	9,940	9,880	9,860	50,000	5
D2	4,600	5,800	9,910	9,930	9,890	9,870	50,000	5
D3	10	10	10	9,860	10	100	10,000	1
D4	10	10	10	4,400	10	5,560	10,000	1
D5	10	10	10	1,400	8,560	10	10,000	1
Totals	14,550	14,530	11,640	35,530	28,350	25,400	130,000	
Seats	1	1	1	4	3	3		13

Since the quotas are all included in the interval between zero and one, a solution that satisfies the quotas requires zero or one seat for each district/list pair. So, the fourth, fifth and sixth lists can receive, at most, 6 seats overall in the first and second district. Hence, the other lists (the first, second and third) must receive at least $10 - 6 = 4$ seats in the same districts. But these lists are entitled to only 3 seats in total! We are forced to conclude that there is no apportionment with 0 or 1 seats. Necessarily, the fourth, fifth or sixth lists must receive 2 seats in the first or second district, thereby violating the quotas.

There is no unique way to approach the Biproportional apportionment problem. There are two main lines of thought. In one approach, some important properties are stated that every method should satisfy, and then a method is found that indeed satisfies them. This is the approach proposed by Balinski and Demange [6, 9, 10]. The other line of thought starts from the definition of ideal quotas, after which an apportionment is found that minimizes the error (measured in some way) with respect to the quotas. A survey of these methods can be found in [80].

Presenting these methods in detail is beyond the scope of this book. Designing an efficient and sound method for the Biproportional apportionment problem requires a solid mathematical competence, and so it must be left to a mathematician. Otherwise, serious problems can arise, as we shall see next. It is not conceivable to describe a procedure within the context of the law. An algorithm, given its complexity, has to be described in a precise way according to a mathematical formalization, and cannot be described with the same precision by using a juridical language, as law articles are normally written.

We will simply sketch how the problem is faced with an example. Let us assume that the problem data are those in Table 12.2 with three districts (1, 2 and 3) and three lists (A, B and C). In the table entries, we report the votes obtained by each list in each district. In the last row, we report the total vote for each list. The seats assigned to each district have been computed before the elections, and are 9, 7 and 4, respectively, for a total of 20 seats (last column). In the example, we have imagined a high abstentionism rate in the second district (indeed, there are too few votes with respect to the allocated seats). With variable abstentionism rates, and, also, with lists

Table 12.2 Votes obtained by lists *A*, *B* and *C* in the districts 1, 2 and 3

	A	B	C	Tot	Seats
1	310	457	667	1434	9
2	445	226	26	697	7
3	264	11	405	680	4
Tot	1019	694	1098	2811	
Seats	7	5	8		20

Table 12.3 Regional quotas for the votes in Table 12.2

	A	B	C	Seats
1	1.946	2.868	4.186	9
2	4.469	2.270	0.261	7
3	1.553	0.065	2.382	4
Seats	7	5	8	20

Table 12.4 Apportionments for the votes in Table 12.2

	A	B	C
1	2	3	4
2	4	2	1
3	1	0	3

By minimizing the deviations

	A	B	C
1	2	2	5
2	4	3	0
3	1	0	3

By divisor methods

that may or may not be present in some districts, problems can arise, if an allocation method is too 'naive'.

From the second to the last row of the table, we compute the seats to be allocated to the lists at the national level (the values 7, 5 and 8 in this simple case are produced by both the Largest remainder method and the main divisor methods). The votes are transformed into regional quotas, and we obtain the Table 12.3.

If we approach the problem by considering each deviation from the quotas as an error, we may think of minimizing the global error that is incurred when we necessarily move away from the quotas by assigning integer seats. There are various ways of defining the global error: we may decide that the global error is the maximum deviation, or the sum of all deviations, or the sum of the squares of all deviations. Alternatively, we might consider a penalty when we round to the 'wrong' integer, i.e., the one most distant from the quota, and we want to minimize the sum of all penalties. There are mathematical methods that find an apportionment efficiently and quickly for each one of these approaches. They may yield the same apportionment, but this is not necessarily so. By applying these methods to the example in Tables 12.2 and 12.3, we obtain the apportionment on the left side of Table 12.4 (all methods give the same apportionment in this case).

Divisor methods operate on both rows and columns in the same way we have mentioned in Sect. 9.4. A technique, known as *Discrete Alternate Scaling*, proposed by F. Pukelsheim [77], finds an apportionment in every row by transforming a table of votes in one of quotas via a multiplication factor R, that is in general different for each row (see the previous Sect. 9.4). If the sums in the columns are not satisfied, the procedure is repeated, this time column by column, in the table just modified. Now, column sums are satisfied, but the row sums have very likely been 'damaged' by the modification. So, the procedure is repeated for the rows. After a few steps, the procedure stabilizes on the correct apportionment (there are rare cases when the procedure does not converge to a solution). Another procedure, proposed by Balinski and Demange [9], and subsequently called *Tie and Transfer*, is based on the idea that, if the quota coincides with the rounding threshold value, one may round either up or down. Then, the method tries to find a quota table with many values on the threshold so that we can move seats at ease. By applying either method, we find the apportionment on the right side of in Table 12.4.

Which solution is the 'right' one? The answer is: both and neither. Both apportionments satisfy some criteria, but do not satisfy others. Lawmakers are entrusted with the issue of deciding which criteria need to be satisfied. Once this is specified and fixed, the 'right' solutions are those that satisfy the criteria.

From this too short and necessarily incomplete description of the methods for computing seat apportionments, one may get the idea that these techniques are by no means within everyone's reach. This observation, however, raises a problem: how to guarantee to the voter that the apportionment does indeed have the properties stated by the law? This is a necessary transparency requirement, both in regard to fixing errors and avoiding ballot rigging. Luckily, it is possible, for the methods that minimize the deviations from the quotas, to add more information to the apportionment, a so-called *certificate*, by which, in a way that is reasonable, simple and within everyone's reach (at least to the people in the House in charge of computing the apportionment), one can *prove* that the obtained solution satisfies the requirements.

Such an idea is presented in [88, 89]. Just to get a hint of how it can work, consider again the previous example on Table 12.1, with six lists and five districts, in which it is proven that no apportionment with zero or one seats exists. Imagine that an apportionment is produced that, in the first district, assigns one seat to the first, fifth and sixth lists and two seats to the fourth list. The second list, which has received its seat in the second district, complains, claiming that its seat should be assigned in the first district, where it has received many more votes. By using the same reasoning as we did previously, it is possible to 'prove' that such a request cannot be fulfilled unless other lists are penalized by a larger amount. One can read, in [89], the imaginary dialog between Dante and Virgil about a hypothetical election in medieval Siena.

12.2   A Simple Case: One Seat per District and Two Parties

It seems natural that, by having one seat per district and only two competing parties, the allocation procedure becomes much simpler and a satisfactory apportionment is available. We are not referring to Italy, which has a tradition of plurinominal districts and a rather long list of parties, but to a different electoral possibility for the U.S., where the uninominal system has historically been employed, i.e., the candidate who has received more votes in the district is elected. We have stressed several times that the uninominal system is quite distortive with respect to the popular vote. The vote percentages can be considerably different from the seat percentages.

For instance, let us imagine the following distribution of votes over five districts of a state for two parties A and B (the numbers have been chosen so as to represent the vote percentages in each district as well):

	1	2	3	4	5	totals
A	55	53	52	51	27	240
B	45	47	48	49	73	260

According to the uninominal system, party A wins in four districts, and so the seat ratio between the two parties is four to one. This would be the seat allocation:

	1	2	3	4	5	totals
A	1	1	1	1	0	4
B	0	0	0	0	1	1

If we look at the popular vote, we see that party B prevails with 52% of votes (and a quota of 2.6) against 48% for party A (and a quota of 2.4). A fair representation (computed by any one of the methods described in Chap. 9) should therefore allocate three seats to B and two to A. A completely different result!

The question of having a 'true' representation in the Congress has been repeatedly debated, but it seems unlikely that such a consolidated habit will be abandoned any time in the near future. In any case, there is no shortage of proposals for improvement. For instance, Balinski [7, 8] proposes a very simple method based on the popular vote for allocating the seats throughout the state. First, a decision is made as to how many seats a party should receive in the whole state based on the popular vote. Then, every party receives its seats in the districts where it received the largest percentage of the vote. The procedure yields the same result as any biproportional apportionment method. With more than two parties, the procedure would not be so simple. In the example, A receives its two seats in the first and second districts, with the seats in the other districts going to B.

	1	2	3	4	5	totals
A	1	1	0	0	0	2
B	0	0	1	1	1	3

Balinski calls this procedure *Fair majority voting*. Essentially, it is an application of the Largest remainder method applied to the regional quotas (the vote percentages, since we have to assign one seat, are the regional quotas exactly). The method is extremely simple, but it has a weak point, as would any method that tries to adapt the logic of the proportional representation of the popular vote to uninominal systems. It is unavoidable that, in some districts, the seat will go to the candidate who has lost in that district. In the example, this happens in the third and fourth districts.

If the population is utterly accustomed to the idea that the winner in the district must be elected, it would be almost impossible to 'swallow' the idea that the 'loser' would end up going to the House. It is rightly observed in [8] that, nowadays, political representation is more important than territorial representation, and the proposal acknowledges this change in perspective. In this respect, see also our observation on Sect. 9.1. Moreover, Balinski [8] remarks that, since there are districts with different populations, it may very well happen that the loser in one district receives more votes than the winner in another district. Why should this 'loser' candidate be less deserving a seat than the other 'winner' candidate?

12.3   The Italian Electoral 'Bug'

We have already mentioned that the Italian way of rounding the quotas follows the logic of the Largest remainder method. Whereas this method works well for simple apportionments without the risk of ending in a deadlock, this is, unfortunately, not true in regard to biproportional apportionments.

The Largest remainder method works by first rounding down the quotas and then assigning the remaining seats where the penalization resulting from the rounding is larger. By applying this idea to the biproportional case, it is very likely that the row and column sums will not match the correct seat numbers.

So, one must move seats from certain district/list pairs to certain other pairs in order to balance the sums. The Italian law prescribes how to carry out this switching operation. However, the problem is complex, and there is no guarantee that the procedure will not go into a deadlock. Not only might the problem be unsolvable, as shown in the previous example in Table 12.1, but even if it is solvable, the procedure is faulty because it may indeed go into a deadlock.

The procedure written into the law [2, 70] starts by assigning the seats district by district according to the Largest remainder rule. In this way the sums for all districts correspond to the required seat numbers. In this phase, some entries in the district/list table have been rounded down and others up.

Now, one checks whether the sum for each list over all districts corresponds to the number of seats assigned to that list at the national level. It almost surely does not. Some lists have extra seats and some others have a dearth of seats. So, one must move seats from the over-represented lists to the under-represented ones. To remove a seat, one looks for the entries in which the rounding was up and, in particular, the entry with the smallest remainder is chosen. Vice versa, we add one seat to the entries in which the rounding was down and, in particular, the entry with the largest remainder is chosen. Just notice that it is taken for granted that such entries exist, but there is no guarantee that this is true.

Clearly, the entry to which a seat has to be taken or added must be identified in a non-ambiguous way. The choice cannot be arbitrary. However, going on in this way through repeated adjustments, it may happen that an entry with the required properties does not exist. Backtracking and modifying previous choices (always according to some precise rule) might involve a prohibitive amount of time spent computing, even on a computer.

If one objects that this deadlock circumstance is 'practically' unlikely, one nonetheless has to admit that in four elections out of six (1996, 2006, 2008 and 2013), the procedure prescribed by the law was not able to find a correct apportionment. So, how was the problem eventually solved? In a very straightforward way: the final apportionment was found *after the original district seat allocation was modified by Presidential Decree*.

One might argue that political representation is more important than territorial representation, as we have already noted after all, and so, by leaving unchanged the seats to which parties are entitled and changing only those pertaining to the districts, no significant harm is caused.

However, this is not the issue. Modifying the seats of a district means altering its representativity. Article 48 of the Italian Constitution, also states that: *Il voto è personale ed eguale, libero e segreto – The vote is personal and equal, free and secret*. The word 'equal' means that, beyond the obvious fact that each vote is counted as one vote, the number of people needed to form a seat must equal throughout the nation (more or less, of course, due to the integrality requirement), so that each citizen 'owns' an 'equal' fraction of a seat. Otherwise, some people will be 'more equal than others' (quoting Orwell).

In 2013, according to the population data from the 2011 census, the four regions of Friuli-Venezia Giulia, Molise, Trentino-Alto Adige and Sardinia were entitled, before the elections, to 13, 3, 11 and 17 seats, respectively. After the elections and the apportionment computations, Friuli-Venezia Giulia and Molise had one seat less, whereas Trentino-Alto Adige and Sardinia had one seat more. If we compute the seat costs before and after the elections, we have the following data:

	FVG	Molise	TAA	Sardinia
before	93,682	103,483	96,623	97,243
after	101,489	155,224	88,571	91,840

In 2006, this switch also involved Trentino-Alto Adige and Molise, always to Molise's detriment. According to the population data from both 2006 and 2013, the seat cost in Molise was almost twice as big as that for Trentino-Alto Adige, or, in other words, the people in Molise are worth half as much as the people in Trentino-Alto Adige. This is a clear violation of Articles 48 and 56 of the Italian Constitution. Strangely enough, no unconstitutionality objection has ever been raised with respect to the cited Presidential Decree. Bruno Simeone, who first spotted this anomaly and has engaged in a deep investigation of it with Aline Pennisi and Federica Ricca [73–75, 93], in which they dubbed it the 'Electoral Bug', has suitably defined this circumstance with the phrase 'One man, half a vote!'.

The most amazing part of this story is the general lack of information and interest. Very few people are aware of this fact, and even the press, who have been contacted many times by Bruno Simeone, have given no weight to the circumstance.

However, things seem to have been changing for the better in recent times. A study has been conducted by Federica Ricca and Andrea Scozzari, together with the Servizio Studi of the House [78], to establish an effective procedure for rounding.

Chapter 13
Political Districting

13.1 Territory Subdivision

In uninominal systems, only one representative per district is elected to the House, the choice being based on the simple majority criterion. We have already stressed the flaws of choosing according to Simple majority when there are at least three alternatives. Here, we deal with another important issue connected to uninominal choices. Whether the elected candidate has received a few votes more than the second choice or has won in a landslide holds little importance for the party that the candidate represents. It is always matter of one seat, which, in the first case, seems to the adversary to have almost been 'stolen' and in the second case, seems almost 'wasted', because so many supporting votes materialized in only one seat.

It is therefore natural that whoever is in charge of designing the districts would try to exploit the demography of a territory to their advantage, with the goal of winning most districts by a small margin.

Before discussing all of the aspects connected to the problem, it may be useful to get an idea by showing a minor example. Take a look at Fig. 13.1. In the large square on top, a territory is represented that has to be divided into three districts. The territory can be seen as being composed by nine smaller squares, with the (expected) votes for two parties A and B being written in each small square. In particular, the votes for A are indicated at the top left and the votes for B at the bottom right in italics. Overall, party A receives 23 votes and party B receives 26 votes.

The three squares below the main graph represent three alternative political districtings of the territory. In each district, the votes gained by the two parties are indicated (normal font for A and italics for B). The votes of the winner are emphasized in boldface.

In the first districting on the left, party B gets its fill of seats by winning every district by a small margin. The advantage of three votes over A is well distributed, one vote for each district, and so B can take all of the seats to the House.

© Springer Nature Switzerland AG 2020
P. Serafini, *Mathematics to the Rescue of Democracy*,
https://doi.org/10.1007/978-3-030-38368-8_13

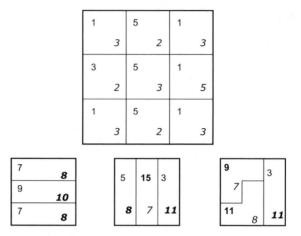

Fig. 13.1 A territory and three possible divisions into three districts

In the second districting (the figure in the center), *B* wins only two seats and *A* is able to win one seat. In two districts, the two parties win by a large margin (8 votes), and in the third district, *B* wins thanks to its global advantage of three votes.

But, if party *A* were in charge of designing the districts, it would choose the third subdivision (figure at right) where it can win two seats out of three. In this case, an overwhelming victory for *B* is confined to one district and, in the other two, it is possible to balance the advantage of 5 votes (the margin of 8 votes for *A* minus the global advantage of 3 votes for *B*) with two margins of 3 and 2 votes.

If one wonders whether it is also possible for *A* to get its fill of seats, the answer is negative. Only the party that has the majority within the whole territory can do that.

Therefore, whoever is in power before the elections will try to redesign the territory in the most favorable way. Of course, this opportunity has a strong conservative effect. Whoever is in power has a greater possibility of remaining in power. In the United States, both the House and the Senate have uninominal districts, and it has been investigated as to whether the opportunity to redesign districts to a single party's advantage has a real impact. According to the results of these studies, the answer seems, indeed, to be affirmative [61, 62].

Moreover, from this example, it is already possible to understand how great a difference there can be between the vote percentages and the seat percentages in a uninominal system. The vote percentages of about 47% and 53% (for *A* and *B*, respectively) can be transformed into 0% and 100% or 33% and 67% or 67% and 33%.

The practice of designing the districts in a 'fancy' way so as to gain the greatest possible advantage is known as *gerrymandering*. The name comes from a merging of the family name of Elbridge Gerry, senator of Massachusetts and later Vice President between 1813 and 1814, and the word 'salamander'. As a senator, Gerry redesigned a district in such a way that, in a satiric drawing, the district was reproduced as a

Massachusetts district, 1812   Uninominal district FVG 2, 2018

Fig. 13.2 Gerrymandering examples

salamander (Fig. 13.2 at left). In this case, the trick did not pay off, because Gerry was not reelected (although this did not prevent him from being elected Vice President immediately afterwards).

Gerrymandering is not a bygone practice. An analysis by the Brennan Center for Justice [83], carried out before the 2018 US mid term election, showed how, because of the gerrymandering effected by the Republicans, the Democrats had to gain a very large margin of the popular vote to win the majority in the House, on the order of 11%. After the election, it could be seen that a margin of 9% was enough to gain the majority. This fact elevated the voices of the people who considered gerrymandering to be irrelevant. This issue was addressed in an editorial in the Wall Street Journal [21]. However, the problem remains important anyway, as observed by several influential thinkers [58].

One can find many fancy maps on the web illustrating gerrymandering, even recent ones. One example, surprisingly similar to the one worked out by Gerry way back in 1812, comes from the 2018 Italian political elections, in which the Friuli-Venezia Giulia region has been divided into five districts, with the second district being every bit as noteworthy as that famous Massachusetts district (Fig. 13.2 at right).

13.2 Equity of a Subdivision

The issue of 'fair' political districting is therefore central and has been an object of study for a long time. A paper from 1965 can be quoted as the first one to try to approach the problem with mathematical tools [50]. It should be emphasized that it was only after the dawn of computers that the problem was approached through the establishment of certain equity criteria that an automatic search could satisfy. Among the many papers that have dealt with the question, we quote a recent survey [44] and an application to political districting in Germany [27]. We also want to mention the paper [28], because it has gained the attention of politicians, so much so that one of the authors was nominated by the Governor of Pennsylvania to serve on a committee for the design of the districts [71].

However, one should not think that computers were immediately greeted as problem solvers in this respect. Rather, they have been looked upon by some as tools for making frauds easier, as argued in the following statement by Justice Harlan of the U.S. Supreme Court (1969): "A computer may grind out district lines which can totally frustrate the popular will on an overwhelming number of critical issues" [62].

And the problem persists even today, as can be seen from an article in the New York Times Magazine from 8/29/2017 [20] that presents the problem and its manipulations in a very detailed way. The title and subtitle are sufficient to show its scope: 'The new front in gerrymandering wars: Democracy versus Math—Sophisticated computer modeling has taken district manipulation to new extremes. To fix this, courts might have to learn how to run the numbers themselves.' Hence, algorithms are welcome as long as they are used by neutral bodies, and not by the politicians themselves.

However, at the moment, the U.S. Supreme Court thinks that the matter is one of political jurisdiction, as ruled on June 27, 2019, by a majority of five against four. In other words, in the Supreme Court's opinion, gerrymandering is a part of normal and acceptable political conflict.

In spite of this opinion, we should face the issue of defining criteria for a fair political districting and also for measuring the fairness degree of a subdivision. A criterion employed in the juridical literature is the so-called *Partisan symmetry* [47, 55, 102]. The criterion corresponds to the idea of anonymity of the parts, as can be read in the definition given in [55]: 'The symmetry standard requires that the electoral system treat similarly-situated parties equally, so that each receives the same fraction of legislative seats for a particular vote percentage as the other party would receive if it had received the same percentage of the vote.'

We must underline that the criterion does not affirm proportionality, but simply party switching. If, for instance, a party gets 70% of the seats against 55% of the votes (which is, in itself, allowed, because we do not pursue proportionality), the other party should also gain 70% of the seats if it receives 55% of the votes (even with a different territorial subdivision, otherwise the criterion would be trivially satisfied).

How to verify that Partisan symmetry is respected is by no means obvious. One way, possibly more understandable, to reformulate the problem, could be the following: if the two parties get the same number of votes, do they receive the same number of seats? Even this reformulation needs further definitions for its implementation. The juridical literature does not provide precise definitions and algorithms that can be immediately implemented. The computation can be based on both hypothetical electoral outcomes and historical data.

In a recent proposal [94], the limits of the concept of Partisan symmetry are outlined and the new concept of the *Efficiency gap* is proposed. This is an interesting idea that evaluates a posteriori the electoral outcome by counting the 'wasted' votes for both parties (the idea can be extended to a greater number of parties, but the presentation in the paper refers to the two-party American system). Votes can be wasted in one of two ways: either because they do not contribute to electing a candidate or because they are beyond the necessary threshold to have a winning candidate. If party A gets 35% of the votes, these are all wasted, whereas party B, which gets 65% of the votes, has wasted the 15% above 50%. The sum of the two wastes is always 50%, as can be easily seen.[1]

The wasted votes for both parties are counted, and one can see how many votes they have wasted in total. Due to the previous observation, the sum of all wasted votes is equal to half of the votes, i.e., a constant. Equity would mean equal waste for both parties, and so, ideally, they should each waste a quarter of all votes. The greater difference between the two values, the less fair the subdivision. The Efficiency gap is defined as this difference (in absolute value) divided by the total number of votes. Hence, it is a number between zero (the parties waste the same number of votes) and 0.5 (in all districts, one party has one vote more than 50%, and so it wastes nothing, whereas the other one wastes all of its votes, i.e. half of the total votes).

We can apply these ideas to the three subdivisions of the example in Fig. 13.1. The subdivision with the smallest Efficiency gap is the second one, and therefore the fairest one according to this concept.

	subd. 1	subd. 2	subd. 3
waste of A	$7 + 9 + 7 = 23$	$5 + 4 + 3 = 12$	$1 + 3 + 1.5 = 5.5$
waste of B	$0.5 + 0.5 + 0.5 = 1.5$	$1.5 + 7 + 4 = 12.5$	$7 + 8 + 4 = 19$
Eff. gap	0.439	0.010	0.276

It can be shown through some simple reasoning that the Efficiency gap, if the populations are equal in all districts, is given by the following formula, where v is the fraction of votes and s is the fraction of seats obtained by one of the two parties (we get the same result if we consider the other party):

$$EG = \left| 2\left(v - \frac{1}{2}\right) - \left(s - \frac{1}{2}\right) \right|$$

The relation can be visualized in Fig. 13.3a, in which the possible seat percentages (Y axis) and vote percentages (X axis) that a party can obtain are shown (we recall that we are dealing with a two-party uninominal system). Not all vote/seat pairs are feasible. The feasible pairs are those within the parallelogram drawn with a heavy line. The square of the percentages is divided into four squares according to a majority/minority of votes and a majority/minority of seats. For instance, 20% of the votes will never produce a majority of seats (more than 50%), because none of

---

[1]Here and subsequently, we simplify the presentation without taking into account that votes are integers and one has to get 50% plus one vote to win.

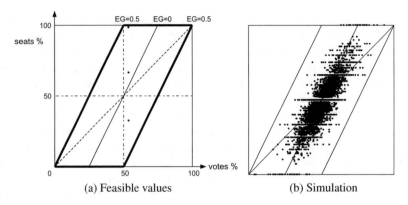

Fig. 13.3 Relationship between the vote and seat percentages and the Efficiency gap

the coordinate points $(20, s)$ with s larger than 50% lie within the parallelogram. Differently, 30% of the votes can produce a majority of 55% seats, because the point $(30, 55)$ lies within the parallelogram (although a majority of 70% of the seats is not possible with 30% of the votes).

The dashed diagonal in the square represents the ideal case of equality between the vote and seat percentages. The oblique solid line corresponds to the case of the zero Efficiency gap. The oblique sides of the parallelogram correspond to the case of the maximum Efficiency gap, equal to 0.5. The other values of the Efficiency gap are obtained on the oblique lines parallel to the one at zero value and the ones at maximum value.

The three points inside of the parallelogram represent the three subdivisions in Fig. 13.1 from the top down. The top one should lie on the parallelogram's side but it has been drawn slightly lower to make it visible.

As can be seen from the previous formula, a zero Efficiency gap does not mean equal percentages between votes and seats. One has an 'amplification' effect of the percentages with respect to the 50% value given by the relation (as can be seen from the formula by putting $EG = 0$):

$$s - \frac{1}{2} = 2 \left(v - \frac{1}{2} \right)$$

This formula expresses the fact that (with a zero Efficiency gap) the percentage of votes above 50% is doubled into the seat percentage beyond 50%. If, for instance, a party receive 60% of the votes, that extra 10% is doubled and results in a seat percentage of 70% (always with a zero Efficiency gap).

In any case, a zero Efficiency gap implies that it is not possible to have a majority of seats in front of a minority of votes and, in general, with an Efficiency Gap of low values, this is an unlikely situation.

By assuming random data and carrying out a probabilistic analysis, an interesting fact emerges: the values of the vote/seat pairs do not get closer to the square diagonal in such a way as to exhibit the same percentages of votes and seats, as one might imagine; rather, they gather near lines with an angular coefficient larger than one, according to the type of random variables. A simulation with 500 points, randomly generated, is shown in Fig. 13.3b, where it can be seen that the points tend to be inclined with angular coefficient 1.5.[2]

We remind the reader that all of these relations are valid within the hypothesis of equal populations. Without this hypothesis, one can imagine paradoxical situations in which a party wins in many districts with very few inhabitants but is totally defeated in large districts. In these cases, one could obtain the majority of seats with minimum vote percentages. Actually, it is very difficult to have districts with exactly equal populations, although it is possible to have comparable populations, and this should be pursued as much as possible. In this case, the relationships are still valid, with good approximation.

13.3   Criteria for Fair Political Districting

Beside these general criteria, many other criteria have been formulated in the literature that should be respected as much as possible [22, 24, 45, 53, 79, 81, 82]. Preliminarily, one has to define minimal units of territory that must not be further split. We call these *territorial units*. The most relevant criteria are:

– *Integrity*: no territorial unit can be split into two or more districts.

– *Contiguity*: the territorial units of a district must be geographically contiguous, i.e., one can move from one unit to another one without leaving the district.

– *Demographic balance*: districts should have more or less the same number of people. In the case of districts with a different number of seats, the cost of a seat should be nearly invariant.

– *Compactness*: each district should be compact.

– *Administrative boundaries*: each district should reproduce administrative borders that are already present in the territory.

– *Natural boundaries*: each district should take care of existing natural borders.

– *Ethnical boundaries*: districts should not split ethnic groups.

The first criterion is quite obvious and corresponds to the idea that territorial units are, so to speak, atomic. Whoever decides which the territorial units are, should design them as sufficiently small in order to prevent 'a priori' manipulations. Also, the Integrity criterion is natural and is usually respected in reality (barring obvious

---

[2]A short note for the mathematicians: one asks the question of what is the expected percentage of seats (over all districts) against a known vote percentage (again, over all districts). The amplification factor of the seat percentage beyond 50% is given by the ratio between the variance of votes and the covariance of vote/seat pairs. The value of this amplification factor varies according to the type of random variables. For uniform variables, it is 1.5. For gaussian variables with a standard deviation of 10% it is almost 4.

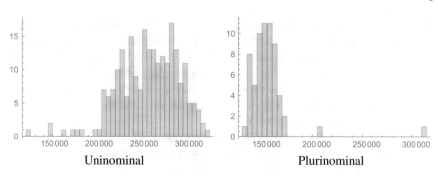

Fig. 13.4 Seat cost distribution—Italian political elections 2018

cases of small islands), because the violation of this criterion can be critically noted by everybody.

The Demographic balance criterion is very important and is directly linked to the principle of one person-one vote. If votes must be equal, representativity must also be equal. Since the exact equality of people for every district is impossible, one has to settle for some percentage bounds within which the population number should fall.

If one looks at how this criterion was respected in the last Italian political elections, one has to admit that the matter is embarrassing. We have already observed (Sect. 9.1) that the cost of a seat for the House of Representatives (Camera dei Deputati) is 96,171 people/seat, data that is obtained by dividing the Italian population (2011 census) by 618 seats (the 12 seats of the Foreign District—Circoscrizione Estero—obviously do not count). The lawmakers have decided to subdivide every district (circoscrizione), both in uninominal and in plurinominal constituencies (collegi), and, as a matter of fact, each voter has, with a single vote, voted for both the uninominal and the plurinominal constituency. Moreover, the number of uninominal seats (Valle d'Aosta has only one uninominal seat and is excluded) has been fixed at 231, while the number of plurinominal seats has been fixed at 386.[3] Hence, the cost of the uninominal seats is almost one and half times higher than that of plurinominal seats (namely, the ratio 386/231 in front of the same population).

This is already a very curious choice. However, beside this question, the variability among the uninominal constituencies of the seat costs is very high. The minimum is given by constituency n. 6 of Trentino-Alto Adige, with 120,413 people and the maximum is given by constituency n. 3, of Sardinia, with 324,470 people, (more than two and a half times the minimum). The histogram in Fig. 13.4 on the left reports the distribution of the populations for the 231 uninominal constituencies. Most values fall within 205,000 and 315,000 people, and one can hardly state that the Demographic balance criterion has been respected.

If we examine the same data for the plurinominal constituencies (Fig. 13.4 on the right), the variability is smaller and is acceptable, except for the anomalous data of

---

[3]Data from Decreto Legislativo, December 12, 2017, n. 189, that established the electoral constituencies.

313,660 people/seat for Molise and 205,895 people/seat for Trentino-Alto Adige. In both cases, the district is made up of a unique constituency, and for Molise, there is only one seat at stake.[4] Hence, Molise was in the strange situation of having two uninominal seats with costs of 149,233 and 164,427 people/seat (much lower than the national average) and a unique plurinominal seat with a cost of 313,660, more or less twice as much.

The Compactness criterion is defined in a vague way. Intuitively, a disk and a square are compact shapes. Those represented in Fig. 13.2 are not. If we want to rely on a computer for a fair district design, we need a precise definition of compactness. Various methods have been proposed, each one with pros and cons. This is not the place to describe how to deal with this problem. It is enough to say that some methods employ the concept of moment of inertia to measure the degree of compactness. Another good index of compactness is represented by the ratio between the area and the square of the perimeter (if we multiply this ratio by 4π we have the so-called Polsby-Popper metric, for which the best compactness value is one and it is achieved by a disk). In any case, mathematical models that have to take compactness into account tend to be complex.

The other criteria that take care of possible administrative, natural (rivers, for instance) or ethnic borders are not universally considered by the literature to be mandatory. The consensus is not unanimous (beside the Administrative boundaries criterion, which is considered the most important). We have to bear in mind that there are perhaps no solutions that can satisfy all criteria. For instance, if we look back at Fig. 13.2 on the right, maybe we can justify the shape of the constituency by thinking of the Ethnic boundaries criterion, in order to keep almost all Slovenian minorities present in the region within the same constituency. However, it is also true that they constitute a small part of the population, and the Compactness criterion should have prevailed.

No matter how complex the mathematical model, this is a problem that only has to be solved every once in a while, and so we can afford a day of computing time to find a satisfactory solution. There exists a vast amount of literature on the subject, and lawmakers should be aware of it without improvising solutions.

Above all, however, an independent body should be entrusted with the political districting problem, because it is unavoidable that the political party in power will find a way, even while respecting all stated criteria, to design the districts to its own advantage. It indeed seems that the majoritarian uninominal system was 'born bad' and that, no matter how one tries to improve it, there will be always malfunctions. We must agree with Balinski's suggestion [8], which we discussed in Sect. 12.2.

---

[4]The fact that there is only one seat in a plurinominal constituency does not mean that this has become a uninominal constituency. Which list gets that seat will depend on the biproportional apportionment of the seats all over the nation, whereas, in the uninominal constituencies, that seat immediately goes to the most voted-for candidate, and consequently to his/her list.

Chapter 14
One Person–One Vote

The principle of 'one person–one vote' is the cornerstone of democracy, and nobody would even consider departing from it. But, is it respected in actual voting systems?

The principle does not simply mean that each vote is counted as one and everybody has the right to vote. We have already observed that, if different districts are entitled to one seat in the parliament but they have different populations, then voters in different districts have different representatives and the principle of 'one person–one vote' is not respected.

Is there a way to 'measure' the voting power of a voter? If it turns out that all voters have the same voting power, we may conclude the the principle of 'one person–one vote' is respected. In addition, we may understand how much voting power a voter has in different voting systems.

14.1 Voting Power

In 1946, Lionel Penrose [76] introduced an index to measure the voting power of a single voter. As may sometimes happen, even to good papers, the result went unnoticed and Penrose's idea remained unknown. In these cases, it is not surprising that, after some time, another person proposes the same idea. This was Banzhaf, who wrote a few papers on the argument [18, 19]. Yet, this was not enough for the idea to spread and be accepted in the scientific community.

As a matter of fact, a few years earlier, the Nobel Prize winner in Economics Lloyd Shapley had written a very influential paper [91] in which he proposed a particular fair share, known, indeed, as *Shapley value*, that should be assigned to each member of a coalition. It turned out that this concept could also be adapted to electoral systems, and Shapley, together with Shubik, wrote a paper [92] that dealt with the computation of the share to be assigned to each voter. This is not exactly voting power, but the paper received a great amount of attention and the Shapley-Shubik index was considered the right tool for measuring voting power, thereby, of course, obscuring Banzhaf's work (not to mention Penrose's).

© Springer Nature Switzerland AG 2020
P. Serafini, *Mathematics to the Rescue of Democracy*,
https://doi.org/10.1007/978-3-030-38368-8_14

This story is well told by Felsenthal and Machover [36, 37], who put Penrose's work in the right perspective showing how the Penrose index (or Penrose-Banzhaf, as it is also called) is the right concept for measuring voting power.

Although the index can be defined for any voting system, its computation is much simpler for binary voting systems, i.e., when each voter can vote either 'yes' or 'no' and the outcome is 'Yes' or 'No' according to some predefined rule.

However, ternary systems do exist in many situations. Very often, abstention is a third possibility in voting and, though, in some systems, only favorable votes count (practically, abstentions are often treated as a 'no' vote), there are other systems that depend in a subtle way on the difference between abstention and a contrary vote. For instance, the UN Security Council has five permanent members and eleven non-permanent members. The rule is that a resolution passes if there are at least nine votes in favor, but no permanent member votes against it. This clause clearly gives a lot of voting power to the permanent members. The computation of non-binary systems tends to be complex and requires some combinatorial mathematics that is beyond the scope of this book. So, we limit ourselves to dealing with binary systems.

It seems safe to consider only those rules that obey the following quite sensible property: in a situation in which the outcome is 'Yes' and a voter has voted 'no', if this voter changes their vote to 'yes', the outcome remains 'Yes'. Symmetrically, in a situation in which the outcome is 'No' and a voter has voted 'yes', if this voter changes their vote to 'no', the outcome remains 'No'.

An almost universally adopted rule that satisfies this property is: the outcome is 'Yes' if and only if the number of 'yes' votes is at least a defined quota q, otherwise it is 'No'. If there are n voters, the unanimity rule corresponds to the case of $q = n$, i.e., all must vote 'yes' for the bill to pass. If $q = \lceil (n + 1)/2 \rceil$ (i.e., $(n + 1)$ divided by two and rounded up), we have the usual majority rule that requires the number of 'yes' votes to be larger than the number of 'no' votes.

The case of two-party voting can be assimilated to the majority rule of 'yes'-'no' voting if n is odd (voting 'yes' is voting for one party and voting 'no' is voting for the other party). If n is even, we should decide what to do if the two parties receive the same number of votes. For 'yes'-'no' votes, this case usually produces a 'No' outcome, unless there are provisions that give a double vote to the president of the assembly in this special case. We can simply neglect the case of equal votes if we assume that n is a large number, as it typically happens in political elections.

The Penrose index can be defined in two equivalent ways. First, we consider all possible partitions of votes into 'yes' and 'no'. Let us call a particular partition of the voters into 'yes' and 'no' votes a voting pattern. With n voters, the total number of voting patterns is 2^n, which is a huge number when n is large. Then, given a particular voter k, we count the number of voting patterns that yield the outcome 'Yes' and in which k has voted 'yes' and the number of voting patterns that yield the outcome 'No' and in which k has voted 'no'. In other words, we count the number of voting patterns in which voter k is on the side of the 'winners'. Let us denote by r the ratio of this number divided by the number of all voting patterns, i.e., 2^n.

Alternatively, we may compute the probability ψ that a particular voter can revert the outcome by changing his/her vote. It can be proved that $\psi = 2r - 1$.[1]

Of the two indices r and ψ defined above, the index ψ seems to be more interesting and significant. Penrose introduces both r and ψ, while Banzhaf later independently defines ψ. We can call ψ the Penrose index (or, if we prefer, the Penrose-Banzhaf index). We note that ψ measures both the absolute voting powers of two different voting systems and the relative voting powers of different voters for a particular voting system.

For instance, with five voters and unanimity rule, there is only one voting pattern that produces a 'Yes' outcome, while all other 31 voting patterns produce a 'No' outcome. Since each voter is in 16 voting patterns with a 'yes' and in another 16 patterns with a 'no', any particular voter is on the side of the winners once ('Yes' outcome) plus 16 times ('No' outcome), for a total of 17 times. So, $r = 17/32$.

For the unanimity rule, a particular voter can reverse the outcome in only two cases: everybody else has voted 'yes' and the voter switches his/her vote from 'yes' to 'no', reversing the outcome from 'Yes' to 'No', or the voter switches his/her vote from 'no' to 'yes', reversing the outcome from 'No' to 'Yes'. So, $\psi = 2/32 = 1/16$.

With five voters and majority rule, if a particular voter has voted 'yes', there must be at least two other 'yes' votes for the outcome to be 'Yes', and this happens 6 times (two 'yes' votes) plus 4 times (three 'yes' votes) plus one time (four 'yes' votes), for a total of 11 voting patterns. If the same voter has voted 'no', there must be at least two 'no' votes for the outcome to be 'No', which again happens 11 times. In total, we have 22 times out of 32 voting patterns and $r = 22/32 = 11/16$.

Also, in this situation, a particular voter k can reverse the outcome in two cases: exactly three voters (including k) have voted 'yes', or exactly two voters (excluding k) have voted 'yes'. Both the first case and the second case happen six times. So, we have $\psi = 12/32 = 3/8$. Hence, voting with majority rule gives a voter a voting power that is six times greater than it would be with unanimity rule. We see that, by using the Penrose index, we can 'measure' the intuitive idea that the unanimity rule is a conservative way of voting.

In general, with n voters and a quota of at least q 'yes' votes for a 'Yes' outcome, we may compute ψ as follows: there are four possible cases according to the vote of the generic voter, 'yes' or 'no', and the general outcome, 'Yes' or 'No'. In the

---

[1]Here, we provide a proof for mathematically-oriented readers. Pick a particular voter. There are four possible cases according to the vote of this voter, 'yes' or 'no', and the general outcome, 'Yes' or 'No'. Let $n_{yY}, n_{yN}, n_{nY}, n_{nN}$ be the number of voting patterns for the four respective cases. Clearly then, $r = (n_{yY} + n_{nN})/(n_{yY} + n_{yN} + n_{nY} + n_{nN})$. Each voting pattern can be paired with another voting pattern in which only the particular voter changes his/her vote. By the stated property, each voting pattern in ('no','Yes') is paired with a voting pattern in ('yes','Yes'). The voting patterns in ('yes','Yes') that are not paired with any voting pattern in ('no','Yes') are necessarily paired with a voting pattern in ('no','No'), i.e., these are exactly those cases in which the voter can reverse the outcome. These voting patterns are, in number, $n_{yY} - n_{nY}$. We can use similar reasoning for the voting patterns in ('yes','No') and find $\psi = (n_{yY} - n_{nY} + n_{nN} - n_{yN})/(n_{yY} + n_{yN} + n_{nY} + n_{nN})$. Now, the thesis follows easily.

cases ('yes','No') and ('no','Yes'), the outcome cannot be reversed. In the case ('yes','Yes'), the outcome can be reversed if and only if there are exactly q 'yes' votes (including the generic voter). In the case ('no','No'), the outcome can be reversed if and only if there are exactly $q-1$ 'yes' votes. Both cases happen $\binom{n-1}{q-1}$ times.[2] Hence, the Penrose index is

$$\psi = \frac{1}{2^{n-1}} \binom{n-1}{q-1}$$

For the majority rule, assuming n to be odd, we have

$$\psi = \frac{1}{2^{n-1}} \binom{n-1}{\frac{n-1}{2}}$$

This index can be approximated when there are many voters by using the Stirling formula[3] as

$$\psi \approx \sqrt{\frac{2}{\pi(n-1)}} \tag{14.1}$$

The interesting fact about this expression is that it decreases as $1/\sqrt{n}$. Penrose has already noted that the number of representatives of groups of different sizes, like the nations in the UN assembly (or, we might add, the members of the European Parliament), should be proportional in number to the square root of the respective populations to give equal voting power to the citizens of the various nations.

This observation is motivated by the idea that the group of representatives of a nation could vote as a block, but this is not necessarily true in general. If we compare the number of seats assigned to the various nations in the European Parliament and their respective populations (Figs. 9.1 and 9.2), we can see that this 'square root rule' is not respected. However, we will later describe a notable example in which the representatives vote as a block, i.e., the US Presidential election.

14.2   Voting Power for Weighted Votes

It is conceivable that voters may have a priori different weights and that the outcome will depend on the sum of the weights, rather than the number of votes. We assume that the weights are positive integers $w_1, \ldots, w_n$ and the outcome is 'Yes' if and only if the sum of the weights of the 'yes' votes is at least a predefined quota q, otherwise it is 'No'.

---

[2] We recall that $\binom{n}{m} = (n!/(m!(n-m)!))$ and the factorial $n!$ is defined as $n! = n \cdot (n-1) \cdot (n-2) \cdots 3 \cdot 2 \cdot 1$.

[3] The Stirling formula approximates the factorial as $n! \approx n^n e^{-n} \sqrt{2\pi n}$.

If a particular voter k with weight w_k has voted 'no' and the outcome is 'No', he/she can reverse the outcome by voting 'yes' if and only if the sum of the weights of the voters who voted 'yes' is a number between $q - w_k$ and $q - 1$ (included). The number must be less than q, otherwise the outcome cannot be 'No', and must be greater than or equal to $q - w_k$, otherwise the outcome cannot be reversed by adding w_k to this number and obtaining at least q.

We need to compute how many voting patterns end up with a sum of weights between $q - w_k$ and $q - 1$. This is not a straightforward task. Let us first give a simple example with four voters and respective weights of $w_1 = 1$, $w_2 = 2$, $w_3 = 2$, $w_4 = 3$. The sum of the weights is 8. So, we fix a quota $q = 5$ corresponding to a majority rule. There are 16 voting patterns which we list below (each voting pattern is numbered and shown as a string of zeros and ones, zero for 'no' and one for 'yes'), together with the sum of the weights of the 'yes' votes.

1 - (0000) - 0	2 - (0001) - 3	3 - (0010) - 2	4 - (0011) - 5
5 - (0100) - 2	6 - (0101) - 5	7 - (0110) - 4	8 - (0111) - 7
9 - (1000) - 1	10 - (1001) - 4	11 - (1010) - 3	12 - (1011) - 6
13 - (1100) - 3	14 - (1101) - 6	15 - (1110) - 5	16 - (1111) - 8

The voting patterns 1, 2, 3, 5, 7, 9, 10, 11 and 13 yield the outcome 'No' (the sum of the weights of the 'yes' votes is less than 5). Let us consider voter n. 1. This voter votes 'no' in the patterns from 1 to 8. So, the patterns in which the outcome is 'No' and voter 1 votes 'no' are 1, 2, 3, 5 and 7. If voter 1 switches his vote, these patterns become patterns 9, 10, 11, 13 and 15. Only pattern 15 has outcome 'Yes'. So, voter 1 can reverse the outcome in only one case.

If we consider the symmetric case (reverting from 'Yes' to 'No'), we get the same result. So, voter 1 can reverse the result two times out of 16 patterns. His Penrose index is $2/16 = 1/8$.

Let us repeat the same computation for the more powerful voter 4. This voter votes 'no' in patterns 1, 3, 5, 7, 9, 11, 13 and 15. So, the patterns in which the outcome is 'No' and voter 4 votes 'no' are 1, 3, 5, 7, 9, 11 and 13. If voter 4 switches her vote, these patterns become patterns 2, 4, 6, 8, 10, 12 and 14. Only patterns 2 and 10 do not reverse the outcome. So, voter 4 can reverse the outcome in five cases. The symmetric case is similar. So, voter 4 has a Penrose index of $10/16 = 5/8$, i.e., five times more than voter 1, in spite of the fact that her weight is three times more! The reader can do the same computation for voters 2 and 3 whose Penrose index is $3/8$.

The procedure that we have outlined for four voters cannot be extended to a larger number of voters because listing all 2^n voting patterns could require prohibitive computing times for large n. For instance, with $n = 58$ (this number is not chosen randomly), assuming that listing one voting pattern together with the related computations takes one nanosecond on a computer, the computation of the Penrose index for all 58 voters according to this naively exhaustive method would require about nine years!

We may wonder whether there exists some easy formula that allows for the computation of the Penrose index for the case of weighted votes. Well, there are theoretical reasons that make such a possibility doubtful. If there were an easily computable formula, then an open problem in theoretical computer science would be solved. This open problem is the famous P versus NP question, which is considered one of the most important and intriguing of unsolved mathematical questions.

However, things are not so bad after all. It is possible to build an algorithm that can compute the Penrose index in reasonable time. The description of said algorithm is beyond the scope of this book. The case of 58 voters can be solved in a few seconds of computing time.

We are talking here of computing the *exact* Penrose index. What about an approximate value based on probabilistic assumptions? This analysis has been carried out [42], and it turns out that, for large n, the Penrose index grows linearly with the weights (we stress that this is not true for a small number of voters or some particular values for the weights).

14.3   The U.S. Presidential Election

As is well known, the U.S. Presidential election is a two-stage election. To each state, plus the District of Columbia, a certain number of electoral votes are assigned based on the populations of the latest census, for a total of 538 electoral votes. At election time, state by state, the candidate who receives more votes in one state receives all of the electoral votes from that state. The candidate who receives at least the quota of 270 electoral votes is elected President.

We must mention that Maine and Nebraska do not assign their electoral votes according to the winner-takes-all rule, but rather according to a method that is a mix of proportional and uninominal. This complicates the computation. However, we can get a very good approximation by virtually splitting each of these states into a number of smaller states equal to the electoral votes, and splitting the population as well. Hence, Maine, which has four electoral votes, is split into four 'copies', each one with a quarter of the population and only one electoral vote. The state of Nebraska has five electoral votes and is split into five copies, each one with a fifth of the population and only one electoral vote. The voting powers of the voters in each copy are equal and correspond to the voting power of any citizen in that state.

Although the U.S. Presidential election is not a binary voting method (as we remarked on Sect. 7.2 third party candidates have participated in presidential elections and their presence very likely altered the final result), for the purpose of computing the Penrose index, we may consider a scenario in which only two candidates are running for President.

We have to ask what the probability is of a single voter reversing the final result by switching his/her vote to the other candidate. Clearly, all voters within a state have the same power, but we may wonder whether voters in different states have different levels of voting power, since the populations are widely different.

One one side, we have to understand what the voting power of a voter in a single state is, i.e., the probability of swinging all electoral votes to the other candidate by switching one vote, and on the other side, we have to compute the probability that, once all of the electoral votes from this state have been switched, the final outcome will be reversed.

We have already computed the Penrose index for the former case in Sect. 14.1 and have seen that it is proportional to $1/\sqrt{n_k}$, with n_k being the population in state k. Actually, n_k should be the number of voters, and we could indeed use this data for an analysis a posteriori of the vote. However, we are interested in an analysis a priori, which should be valid in any circumstance, and therefore we have to consider all possible voting patterns as being equally likely. For this analysis, we necessarily have to take into consideration the population data. If, as we may expect, the fraction of the population that goes to the polls is almost the same in each state, we may safely consider n_k to be the population. Proportionality among the various states is left invariant.

So, it seems that voters in larger states are less powerful, because of the term $1/\sqrt{n_k}$, which gets smaller as n_k increases (four times the population means half the voting power). Indeed, they are less powerful in their ability to reverse the outcome of their own states, but larger states have a larger number of electoral votes, and thus can reverse the final outcome more easily than smaller states.

This second stage of the Presidential election, when we have to sum the electoral votes of the various states, is just a weighted voting system with weights that are equal to the electoral votes. We have said that, for a weighted voting system, the voting power is approximately proportional to the weights, i.e., to the electoral votes. In turn, these are almost proportional to the populations.

Combining the two results, we find that the approximate Penrose index for the voters of state k is proportional to $\sqrt{n_k}$. This time, the proportionality is direct and not inverse, so that a voter in a larger state has more voting power than a voter in a smaller state.

We may compute the Penrose index exactly by using the algorithm cited in the previous section. In Table 14.1, we show the result of this computation for all 50 states and the District of Columbia. In the table, we report the population data from the 2010 census, the electoral data for the next presidential election in 2020 and the Penrose index normalized so that the smallest index (for Montana) is equal to 1. As is apparent from the table, a voter in California has 3.394 more in voting power than a voter in Montana. We can hardly consider this voting system to be consistent with the principle of one person–one vote.

If the election were to be based on the popular vote, then all voters would share the same voting power. It is instructive to compare the voting power for this type of election to the values listed in Table 14.1. Recall that, in this table, the values are normalized. Their absolute values can be obtained by multiplying each entry by $1.81461 \cdot 10^{-5}$ (so this is the absolute value for Montana). To compute the Penrose index for an election based on the popular vote, we just have to apply the formula (14.1) with $n = 308\ 745\ 538$, which is the entire U.S. population. The result is $4.54088 \cdot 10^{-5}$. If we divide this number by $1.81461 \cdot 10^{-5}$, we get 2.50239, a value

Table 14.1 2020 US presidential election: populations, electoral votes and normalized Penrose indices

State	pop.	e.v.	index	State	pop.	e.v.	index
AL	4 779 736	9	2.493	AK	710 231	3	2.152
AZ	6 392 017	11	2.638	AR	2 915 918	6	2.126
CA	37 253 956	55	6.191	CO	5 029 196	9	2.431
CT	3 574 097	7	2.241	DE	897 934	3	1.914
DC	601 723	3	2.338	FL	18 801 310	29	4.141
GA	9 687 653	16	3.129	HI	1 360 301	4	2.074
ID	1 567 582	4	1.932	IL	12 830 632	20	3.412
IN	6 483 802	11	2.619	IA	3 046 355	6	2.080
KS	2 853 118	6	2.149	KY	4 339 367	8	2.325
LA	4 533 372	8	2.275	ME	1 328 361	4	1.049
MD	5 773 552	10	2.522	MA	6 547 629	11	2.607
MI	9 883 640	16	3.098	MN	5 303 925	10	2.631
MS	2 967 297	6	2.107	MO	5 988 927	10	2.476
MT	989 415	3	1.823	NE	1 826 341	5	1.000
NV	2 700 551	6	2.209	NH	1 316 470	4	2.108
NJ	8 791 894	14	2.869	NM	2 059 179	5	2.107
NY	19 378 102	29	4.079	NC	9 535 483	15	2.954
ND	672 591	3	2.211	OH	11 536 504	18	3.232
OK	3 751 351	7	2.187	OR	3 831 074	7	1.187
PA	12 702 379	20	3.429	RI	1 052 567	4	2.357
SC	4 625 364	9	2.535	SD	814 180	3	2.010
TN	6 346 105	11	2.648	TX	25 145 561	38	4.778
UT	2 763 885	6	2.183	VT	625 741	3	2.293
VA	8 001 024	13	2.790	WA	6 724 540	12	2.808
WV	1 852 994	5	2.221	WI	5 686 986	10	2.541
WY	563 626	3	2.416				

that can now be directly compared to the values in Table 14.1. In only two states do the voters have a voting power larger than they do with a popular vote, namely, California and Texas. All other states have smaller values.

We may compute the Penrose index exactly by using the algorithm cited in the previous section. In Table 14.1, we show the result of this computation for all 50 states and the District of Columbia (due to the previous observation about Maine and Nebraska, the computation is actually carried out for 58 'states'). In the table, we report the population data from the 2010 census, the electoral data for the next Presidential election in 2020 and the Penrose index normalized so that the smallest index (for Nebraska) is equal to 1. As is apparent from the table, a voter in California has 6.191 more in voting power than a voter in Montana. We can hardly consider this voting system to be consistent with the principle of one person–one vote.

An interesting observation can be made in regard to the method in use in Maine and Nebraska. If they were to use the winner-takes-all rule, they would have roughly twice as much voting power as they do with the current method. This result is consistent with the observation that the voting power is proportional to $\sqrt{n_k}$. If n_k is divided by four (as for Maine), the voting power is reduced by half.

If the U.S. Presidential election were to be based on the popular vote, then all voters would share the same voting power. It is instructive to compare the voting power for this type of election to the values listed in Table 14.1. Recall that, in this table, the values are normalized. Their absolute values can be obtained by multiplying each entry by $9.96591 \cdot 10^{-6}$ (this is the absolute value for Nebraska, indeed). To compute the Penrose index for an election based on the popular vote, we just have to apply the formula (14.1) with $n = 308\,745\,538$, which is the entire U.S. population. The result is $4.54088 \cdot 10^{-5}$. If we divide this number by $9.96591 \cdot 10^{-6}$, we get 4.55641, a value that can now be directly compared to the values in Table 14.1. In only two states do the voters have a voting power larger than they do with a popular vote, namely, California and Texas. All other states have smaller values.

However, some readers may consider these findings to be almost meaningless in practice. Everybody knows that, in states like California or Texas, the result of a Presidential election is quite predictable, while in the so-called swing states, the votes count much more, and so the voters are actually more powerful (thus contradicting the previous observations). This was, for instance, the position taken by Margolis [63], who considered the Banzhaf (and, indirectly, the Penrose) index 'fallacious', because it does not take into account actual demographic or voting patterns. We remark that these observations are more pertinent for a study of a particular election outcome, but, if we have to decide what voting system has to be chosen, then this system must work with any combination of voting patterns, present and future.

Chapter 15
Some Additional (and Personal) Considerations

Everything that has been written in the previous chapters concerns the practice of voting, and the main goal has been that of presenting the theoretical and technical aspects that are at the foundation of the electoral mechanisms. These are important and useful notions if, of course, people do vote. But will people still vote in the more or less near future? After all, it was not that long ago that, in some places, people were not voting much at all and nothing precludes such times occurring again.

This concern should not be overlooked. Some recent electoral outcomes, in particular, in the United States, though the phenomenon is transversal, have marked a turning point in the very concept of voting that may yet lead us in a direction that is unclear.

It has always been stressed that democracy needs information, knowledge, and awareness in order to function well. This is certainly true. If, in older time, people did not vote, this was also because the ordinary citizen was not considered sufficiently aware and worth of voting. Tomorrow, a similar theory could be again affirmed in the face of increased complexity within the structures of a society, of a nation and of the interrelations between nations.

We believe that the fundamental point of every consideration is the fact that the only person who is entitled to represent the interests of an individual is precisely that individual, and nobody else. No matter how uninformed he/she can be, his/her interests cannot be represented by those few who are 'in the know' and are 'worthy' of voting. These are human beings and, as such, they will pursue their own interests first and foremost. Perhaps they will also take into consideration the interests of that individual who does not vote, but that will happen accidentally.

So, how to balance the need to allow everyone to vote with the unavoidable fact that not everybody has enough knowledge to understand how the state and the society function?

In our opinion, the answer lies in those interfaces between citizens and the institutions that were called parties. We use the verb in the past tense because, nowadays, parties seem different from those that came before. Historically, parties elaborated the demands, concerns and all various requests that came from the society and turned

© Springer Nature Switzerland AG 2020
P. Serafini, *Mathematics to the Rescue of Democracy*,
https://doi.org/10.1007/978-3-030-38368-8_15

them into political projects. The 'knowledge' played a role in this phase, filtering the possibly irrational and diverging drives of the society by making them coherent and viable.

As far as the role of the parties is concerned, and of the intermediate bodies of the society in general, we must add some observations that we consider important and that, perhaps, have not been taken properly into consideration. These observations concern the role of Adam Smith's so-called 'invisible hand'. As is well known, the fundamental idea of economic liberalism is based on the assumption that the actions of individuals, even if they are motivated by their own interests, will, once they are left free to their own devices, also act in the interest of the collective welfare, as if they were indeed guided by an invisible hand. No doubt, the history of the last three centuries has shown that the invisible hand has actually operated with great success. But, up to what point does its beneficial action arrive?

In 1951, John Nash[1] developed Non-Cooperative Game Theory [68], which studies the equilibria reached by a society in which individuals act by improving their own utility. Given an initial situation for a society, it may happen that some individuals have an objective interest to take actions that move them away from that situation. This general movement leads the society into another situation from which nobody has any interest in moving away. Hence, this is a stable equilibrium point, known, indeed, as *Nash equilibrium.*

The surprising result is that, even if all individuals act with only their personal advantage in mind thus striving to rationally improve their own position, it may happen that, eventually, they all find themselves in a worse situation than the one they started from. In this case, we say that this is a *dominated equilibrium.*

This seems to be a paradoxical outcome, but dominated equilibria occur much more often than one might think and examples are abundant. To illustrate this situation, Tucker [98] introduced a celebrated paradigm known as the *Prisoner's dilemma.* It is useful to review the paradigm here for the benefit of those who do not know it. Two suspected criminals are separately questioned by the police. There is insufficient evidence to charge them for their crimes, so that, if they keep their mouths shut they will only be charged with a minor crime (like using force against a public official). However, the police try to convince them, separately, to confess, in order to earn some reduction in their sentence for cooperating. The temptation is strong, and, eventually, they both give in, with the result being that each one's sentence reduction is eliminated by the other's disclosure and they both go to jail with a longer sentence than the one that they would have gotten by keeping their mouths shut.

The moral of the paradigm is, indeed, that the rational behavior of pursuing one's own interest can lead to a situation in which everybody is worse off than before. However, it does not seem that this paradigm has been assimilated into common wisdom, in part because the paradigm's content is, perhaps, ill-advised. It would not be at all unusual for someone who reads this story to react with satisfaction—two criminals go to jail, right on!—and miss the principal point of the paradigm, which is negative.

---

[1]John Nash (1928–2015) was awarded the Nobel prize in 1994 for these studies. He is known to a wider audience thanks to the movie 'A Beautiful Mind'.

The arms race is a typical case of the Prisoner's dilemma. Two nations could have a limited armament, but if one of them increases its armament, it gets an advantage over the other one. So, they both continue to increase their stores of arms, finding themselves in a situation exactly like the initial one, but with a much bigger expenditure in armaments. Another example is tax evasion: if everybody pays taxes, a single individual gains an undeniable advantage by not paying taxes. So, everybody tries to evade taxes, eventually placing them in a worsened situation, because society cannot work without fiscal revenues.

The examples could continue. Rational behaviors often lead, almost automatically, to worsened situations. European politics in recent years has often given the impression of proceeding towards an equilibrium point at which each nation and Europe itself will lose.

If there is no higher institution that dictates behavior, it is almost unavoidable that a dominated equilibrium will be reached. In the Prisoner's dilemma, it quite often happens that the rules of organized crime violently sanction whoever confesses and the almost certain punishment leads to a different equilibrium point. Also, in regard to tax evasion, a sanction is foreseen. But there is no higher institution to control arms race or international politics.

Therefore, the Prisoner's dilemma shows that the invisible hand does not necessarily operate for the common welfare. If totally left to itself, it can produce situations in which everybody is worse off. In the literature, the loss in collective utility that is incurred by leaving everybody free to act with respect to an action governed by a superior institution is called the *price of anarchy* [29, 40, 72].

Let us cast these ideas in the social context of voting. We said at the outset that the only person with a legitimate right to represent the interests of an individual is exactly that individual. If the invisible hand were perfect, a direct democracy, as some people are proposing today, would also lead the society toward a generalized welfare. However, we have stressed that the invisible hand has limitations and that a selfish behavior can lead to undesirable situations. So, how should we operate in order to avoid the traps of the dominated Nash equilibria? This could be the role of the parties and, in general, of all intermediate bodies of the society (the importance of the intermediate bodies was already stressed by Alexis de Tocqueville [33]).

Unfortunately, politics has evolved in such a way that parties have become far less reliable institutions. This not only happened in Italy, where the rate of corruption of the parties has been quite high, but also in other equivalent democracies. It does not make much sense to pin the blame on politicians alone. The political class is a mirror of the country in which it lives. The whole society should become ethically aware, for the parties to regenerate their fundamental role.

We add one more observation that concerns the possibility of an electoral democracy surviving. The problem could be put this way: the wealth that has to be divided up must be of sufficient quantity if we want democracy to function. Indeed, a minority of people will accept being governed by an elected majority only if it can enjoy at least some degree of social welfare. If it is left with only crumbs, or in any way perceives what it is getting as insufficient, it will not be open to being governed and so it will seek power by other means. If a nation is poor, it is very likely that it will calcify into a non-democratic situation in which a few rich people hold the many poor people in subjection through force. Countries that are in this condition have a very difficult time evolving toward democracy.

We must not forget that democracy, as we know it today, has slowly evolved in Europe over the last three to four centuries, hand in hand with an increase in welfare, which was, in turn, facilitated by industrialization and the general exploitation of resources in the rest of the world. Nowadays, globalization is rebalancing welfare all over the world. We have enjoyed a good welfare, and it seems unavoidable that, due to rebalancing, we will eventually be worse off, barring some social innovation that could also happen. However, a decrease in wealth undermines democracy, and this is a danger that we have to face.

One step that could be taken, among many, to overcome the difficulties that democracy experiences today in western countries, could certainly consist in improving the electoral systems so that citizens are more aware and active in politics. In this book, some suggestions have been made toward this goal. It may be surprising that mathematics can come to the rescue of democracy. But, if this is understood and accepted, we believe that there can be positive outcomes.

References

1. https://mitpress.mit.edu/books/majority-judgment
2. Legge 4 agosto 1993 n 277. Nuove norme per l'elezione della Camera dei Deputati
3. Arrow KJ (1950) A difficulty in the concept of social welfare. J Polit Econ 58(4):328–346
4. Arrow KJ (1951) Social choice and individual values. Wiley, New York
5. Arrow KJ (1963) Social choice and individual values. Yale University, New Haven
6. Balinski ML (2006) Apportionment: Uni-and bi-dimensional. Mathematics and democracy. Springer, Berlin, pp 43–53
7. Balinski ML (2008) Fair majority voting (or how to eliminate gerrymandering). Am Math Mon 115(2):97–113
8. Balinski ML (2014) How to make the house of representatives representative. In: The conversation. https://theconversation.com/how-to-make-the-house-of-representatives-representative-32921
9. Balinski ML, Demange G (1989) Algorithms for proportional matrices in reals and integers. Math Program 45(1–3):193–210
10. Balinski ML, Demange G (1989) An axiomatic approach to proportionality between matrices. Math Oper Res 14(4):700–719
11. Balinski ML, Jennings A, Laraki R (2009) Monotonic incompatibility between electing and ranking. Econ Lett 105(2):145–147
12. Balinski ML, Laraki R (2010) Majority judgment: measuring, ranking, and electing. MIT Press, Cambridge
13. Balinski ML, Laraki R (2012) Jugement majoritaire versus vote majoritaire. Revue française d'économie 27(4):11–44
14. Balinski ML, Laraki R (2013) How best to rank wines: majority judgment. Wine economics. Springer, Berlin, pp 149–172
15. Balinski ML, Laraki R (2014) Judge: Don't vote!. Oper Res 62(3):483–511
16. Balinski ML, Laraki R (2018) Majority judgement versus approval voting. Oper Res
17. Balinski ML, Young HP (2010) Fair representation: meeting the ideal of one man, one vote. Brookings Institution Press, 2010
18. Banzhaf JF III (1964) Weighted voting doesn't work: a mathematical analysis. Rutgers L Rev 19:317
19. Banzhaf JF III (1965) Multi-member electoral districts-do they violate the one man, one vote principle. Yale LJ 75:1309
20. Bazelon E (2017) The new front in gerrymandering wars: democracy versus math. The New York Times Magazine, 29/8/2017. https://www.nytimes.com/2017/08/29/magazine/the-new-front-in-the-gerrymandering-wars-democracy-vs-math.html

21. Editorial Board (2018) The gerrymander excuse implodes. Wall Street J 16/11/2018. https://www.wsj.com/articles/the-gerrymander-excuse-implodes-1542412885
22. Bodin LD (1973) A districting experiment with a clustering algorithm. Ann N Y Acad Sci 219(1):209–214
23. Borda JC (1784) Mémoire sur les élections au scrutin. Histoire de l'Académie Royale des Sciences, pp 657–665
24. Bozkaya B, Erkut E, Laporte G (2003) A tabu search heuristic and adaptive memory procedure for political districting. Eur J Oper Res 144(1):12–26
25. Brams SJ, Fishburn PC (1978) Approval voting. Am Polit Sci Rev 72(3):831–847
26. Brams SJ, Fishburn PC (2007) Approval voting. Springer Science & Business Media, Berlin
27. Brieden A, Gritzmann P, Klemm F (2017) Constrained clustering via diagrams: A unified theory and its application to electoral district design. Eur J Oper Res 263(1):18–34
28. Chikina M, Frieze A, Pegden W (2017) Assessing significance in a Markov chain without mixing. Proc Natl Acad Sci 114(11):2860–2864
29. Christodoulou G, Koutsoupias E (2005) The price of anarchy of finite congestion games. In: Proceedings of the thirty-seventh annual ACM symposium on theory of computing. ACM, pp 67–73
30. de Condorcet JAC (1785) Essai sur l'application de l'analyse à la probabilité des décisions à la pluralité des voix. l'Imprimerie royale
31. Coombs CH (1964) A theory of data. Wiley, New Jersey
32. Copeland AH (1951) A reasonable social welfare function. Technical report, Seminar on Mathematics in Social Sciences, University of Michigan
33. De Tocqueville A (2003) Democracy in America. Regnery Publishing, Washington
34. Dodgson C, (1876) A method of taking votes on more than two issues. Pamphlet. Reprinted. In: D. Black, (1958) The theory of Committees and Elections. Cambridge University Press, UK
35. Faliszewski P, Skowron P, Slinko A, Talmon N (2017) Multiwinner voting: a new challenge for social choice theory. In: Endriss U (ed) Trends in computational social choice, Lulu.com, p 27
36. Felsenthal DS, Machover M (1998) The measurement of voting power, Edward Elgar Publishing
37. Felsenthal DS, Machover M (2005) Voting power measurement: a story of misreinvention. Soc Choice Welf 25(2–3):485–506
38. FINA (2017) Fina diving rules, 12–23. https://www.fina.org
39. Fishburn PC, Brams SJ (1983) Paradoxes of preferential voting. Math Mag 56(4):207–214
40. Gairing M, Lücking T, Monien B, Tiemann K (2005) Nash equilibria, the price of anarchy and the fully mixed Nash equilibrium conjecture. International colloquium on automata, languages, and programming. Springer, Berlin, pp 51–65
41. Galton F (1907) One vote, one value. Nature 75(1948):414
42. Gelman A, Katz JN, Tuerlinckx F (2002) The mathematics and statistics of voting power. Stat Sci 420–435
43. Gibbard A (1973) Manipulation of voting schemes: a general result. J Econ Soc, Econ, pp 587–601
44. Goderbauer S, Winandy J (2018) Political districting problem: literature review and discussion with regard to federal elections in Germany. https://www.or.rwth-aachen.de/files/research/repORt/LitSurvey_PoliticalDistricting__Goderbauer_Winandy_20181024.pdf
45. Di Cortona PG, Manzi C, Pennisi A, Ricca F, Simeone B (1999) Evaluation and optimization of electoral systems, Society for industrial and applied mathematics (SIAM)., Philadelphia PA
46. Grimmett G, Laslier JF, Pukelsheim F, Gonzalez VR, Rose R, Słomczyński W, Zachariasen M, Zyczkowski K (2011) The allocation between the EU member states of the seats in the European parliament Cambridge compromise
47. Grofman B, King G (2007) The future of partisan symmetry as a judicial test for partisan gerrymandering after LULAC v Perry. Election Law J 6(1):2–35

48. Hägele G, Pukelsheim F (2001) Llull's writings on electoral systems. Stud Lul 41(97):3–38
49. Hägele G, Pukelsheim F (2008) The electoral systems of Nicolaus Cusanus in the catholic concordance and beyond. In: Christianson G, Izbicki TM, Bellitto CM (eds) The Church, the councils and reform: lessons from the fifteenth century. Catholic University of America Press, Washington, DC, pp 229–249
50. Hess SW, Weaver JB, Siegfeldt HJ, Whelan JN, Zitlau PA (1965) Nonpartisan political redistricting by computer. Oper Res 13(6):998–1006
51. Huntington EV (1921) The mathematical theory of the apportionment of representatives. Proc Natl Acad Sci 7(4):123–127
52. Jefferson T (1904) The works of Thomas Jefferson, vol VI, GP Putnam's Sons
53. Kalcsics J, Nickel S, Schröder M (2005) Towards a unified territorial design approach-applications, algorithms and GIS integration. Top 13(1):1–56
54. Kemeny JG (1959) Mathematics without numbers. Daedalus 88(4):577–591
55. King G (2005) Brief of amici curiae professors Gary King, Bernard Grofman, Andrew Gelman, and Jonathan Katz in support of neither party. US Supreme Court in Jackson v. Perry. Amicus Brief 130. http://j.mp/2gw1W1R
56. Krantz D, Luce D, Suppes P, Tversky A (1971) Foundations of measurement, vol I, Additive and polynomial representations
57. Laslier JF (2012) Introduction to the special issue around the cambridge compromise: apportionment in theory and practice. Math Soc Sci 63(2):65–67. Around the Cambridge Compromise, Apportionment in Theory and Practice
58. Leonhardt D (2019) Trump's overhyped speech. New York times, 08/01/2019. https://www.nytimes.com/2019/01/08/opinion/trump-address-border-wall-shutdown.html?rref=collect ion%2Fbyline%2Fdavid-leonhardt&action=click&contentCollection=undefined&region= stream&module=stream_unit&version=latest&contentPlacement=1&pgtype=collection
59. Lippmann W (1993) The phantom public. Originally published by MacMillan, New York, Transaction Publishers, New Brunswick (USA), London (UK), p 1925
60. Maniquet F, Mongin P (2015) Approval voting and Arrow's impossibility theorem. Soc Choice Welf 44(3):519–532
61. Mann TE (2007) Polarizing the house of representatives: how much does gerrymandering matter? Red and blue nation, pp 263–283
62. Mann TE, Cain BE (2008) Party lines: competition, partisanship, and congressional redistricting. Brookings Institution Press
63. Margolis H (1983) The Banzhaf fallacy. Am J Polit Sci 27(2):321–326
64. May KO (1952) A set of independent necessary and sufficient conditions for simple majority decision. J Econ Soc, Econ, pp 680–684
65. Meir R (2017) Iterative voting. In: Endriss U (ed) Trends in computational social choice, p 69. Lulu.com
66. Miller GA (1956) The magical number seven, plus or minus two: some limits on our capacity for processing information. Psychol Rev 63(2):81
67. Moulin H (1988) Condorcet's principle implies the no show paradox. J Econ Theory 45(1):53–64
68. Nash J (1951) Non-cooperative games. Ann Math 286–295
69. Ostrogorski M (1903) La démocratie et l'organisation des partis politiques, vol 2, Calmann-Lévy
70. Ddl Camera 2620 13 ottobre (2005) Modifiche per l'elezione della Camera dei Deputati e del Senato della Repubblica
71. Panko B (2018) Mathematical sciences professor appointed to state commission on redistricting. MCS News, 8/12/2018. https://www.cmu.edu/mcs/news-events/2018/1208_Pegden-State-Commission.html
72. Papadimitriou CH (2001) Algorithms, games, and the internet. International colloquium on automata, languages, and programming. Springer, Berlin, pp 1–3
73. Pennisi A (2006) The italian bug: a flawed procedure for bi-proportional seat allocation. Mathematics and democracy. Springer, Berlin, pp 151–165

74. Pennisi A, Ricca F, Simeone B (2005) Malfunzionamenti dell'allocazione biproporzionale di seggi nella riforma elettorale italiana. Dipartimento di Statistica, Probabilità e Statistiche Applicate, Università La Sapienza, Roma, Serie A-Ricerche, p 21

75. Pennisi A, Simeone B, Ricca F (2006) Bachi e buchi della legge elettorale italiana nell'allocazione biproporzionale di seggi. Sociologia e ricerca sociale. Fascicolo 79:1000–1022

76. Penrose LS (1946) The elementary statistics of majority voting. J R Stat Soc 109(1):53–57

77. Pukelsheim F (2004) BAZI - a Java program for proportional representation. Oberwolfach reports 1:735–737

78. Ricca F, Scozzari A (2019) L'algoritmo elettorale tra rappresentanza politica e rappresentanza territoriale. Roma, aprile, Technical report, Camera dei Deputati, Servizio Studi, XVIII Legislatura

79. Ricca F, Scozzari A, Serafini P (2017) A guided tour of the mathematics of seat allocation and political districting. In: Endriss U (ed) Trends in computational social choice, p 49. Lulu.com

80. Ricca F, Scozzari A, Serafini P, Simeone B (2012) Error minimization methods in biproportional apportionment. Top 20(3):547–577

81. Ricca F, Scozzari A, Simeone B (2013) Political districting: from classical models to recent approaches. Ann Oper Res 204:271–299

82. Ricca F, Simeone B (1997) Political redistricting: traps, criteria, algorithms, and trade-offs. Ric Oper 27:81–119

83. Royden L, Li M, Rudensky Y (2018) Extreme gerrymandering and the 2018 midterm. Brennan center for justice. https://www.brennancenter.org/sites/default/files/publications/extreme%20gerrymandering_2.pdf

84. Satterthwaite MA (1975) Strategy-proofness and arrow's conditions: existence and correspondence theorems for voting procedures and social welfare functions. J Econ Theory 10(2):187–217

85. Schulze M (2003) A new monotonic and clone-independent single-winner election method. Voting Matters 17(1):9–19

86. Schulze M (2011) A new monotonic, clone-independent, reversal symmetric, and condorcet-consistent single-winner election method. Soc Choice Welf 36(2):267–303

87. Serafini P (2012) Allocation of the EU Parliament seats via integer linear programming and revised quotas. Math Soc Sci 63(2):107–113

88. Serafini P (2015) Certificates of optimality for minimum norm biproportional apportionments. Soc Choice Welf 44(1):1–12

89. Serafini P, Simeone B (2012) Certificates of optimality: the third way to biproportional apportionment. Soc Choice Welf 38(2):247–268

90. Serafini P, Simeone B (2012) Parametric maximum flow methods for minimax approximation of target quotas in biproportional apportionment. Networks 59(2):191–208

91. Shapley LS (1953) A value for n-person games. Contrib Theory Games 2(28):307–317

92. Shapley LS, Shubik M (1954) A method for evaluating the distribution of power in a committee system. Am Polit Sci Rev 48(3):787–792

93. Simeone B, Pennisi A, Ricca F (2009) Una legge elettorale sistematicamente erronea. Polena 2(2):65–72

94. Stephanopoulos NO, McGhee EM (2015) Partisan gerrymandering and the efficiency gap. Univ Chic Law Rev 831–900

95. Szpiro G (2010) Numbers rule: the vexing mathematics of democracy, from Plato to present. Princeton University Press

96. Thrasher M, Borisyuk G, Rallings C, Johnston R (2011) Electoral bias at the 2010 general election: evaluating its extent in a three-party system. J Elections, Public Opin Parties 21(2):279–294

97. Tideman N (2017) Collective decisions and voting: the potential for public choice. Routledge

98. Tucker AW (1950) A two-person dilemma. Readings in games and information, pp 7–8

99. US Census. https://www.census.gov/library/publications/1793/dec/number-of-persons.html

100. US Census (2011) https://www.census.gov/prod/cen2010/briefs/c2010br-01.pdf

101. US Census (2018) https://en.wikipedia.org/wiki/1790_United_States_Census
102. Wang SS-H (2016) Three tests for practical evaluation of partisan gerrymandering. Stan L Rev 68:1263
103. Washington G (1939) The Writings of George Washington, vol 32, March 10, 1792–June 30, 1793. United States Government Printing Office
104. Young HP (1988) Condorcet's theory of voting. Am Polit Sci Rev 82(4):1231–1244

Index

A
Adams, John Quincy, 85
Adams' method, 78, 85
Administrative boundaries, criterion of, 109
Alabama paradox, 74, 85
Anonymity of collective choice functions, 41, 42
Approval voting, 39, 54, 57, 92
Arithmetic mean, 23, 24, 78
Arrow, Kenneth Joseph, 6, 35
Arrow's theorem, 38, 64
Aumann, Robert, 6

B
Balinski, Jennings and Laraki theorem, 45
Balinski, Michel, 47
Bayrou, François, 58
Biproportional apportionment, 93
Borda, **21**
    and Condorcet table, 22
    method, 21, 44, 59
Borda, Jean-Charles, 21
Brexit, 19, 81
Bug, electoral, **100, 101**

C
Certificate for seat apportionment, 98
Choice monotonicity, 45, 67
Compactness, criterion of, 109, 111
Condorcet, **11**
    criterion, 14, 64
    cycle, 16
    method, 39, 44, 59
    table, 15
    winner, 14, 59, 64

Condorcet, Marquis de, 11
Constitution
    Italy, 4, 73, 101, 102
    USA, 83
Contiguity, criterion of territorial, 109
Coombs method, 33
Copeland method, 16, 20
Cusanus, Nicolaus, 21
Cyclic preferences, 14

D
Dean's method, 78
Degressive proportionality, 71, 81
Demographic balance, criterion of, 109, 110
de Tocqueville, Alexis, 12, 125
D'Hondt method, 80
Dictatorship, Non-, 37, 64
Direct democracy, 125
Discrete Alternating Scaling, 98
Divisor methods, **76**
Droop quota, 90

E
Efficiency gap, 106
Ethnical boundaries, criterion of, 109
European Parliament, 81, 116
Executive power, 3

F
Fair majority voting, 100, 101
Fair share quotas, 95
Formula One, 8, 24, 62
French elections, 56
Friuli-Venezia Giulia, 101, 102, 105

© Springer Nature Switzerland AG 2020   133
P. Serafini, *Mathematics to the Rescue of Democracy*,
https://doi.org/10.1007/978-3-030-38368-8

G
Galton, Francis, 48
Geometric mean, 78
Gerry, Elbridge, 104
Gerrymandering, 104
Gibbard and Satterthwaite theorem, 42
Giro d'Italia, 9

H
Hamilton, Alexander, 84
Hamilton, Lewis, 24
Hamilton's method, 73, 84
Hamlet, 25
Harmonic mean, 78
Historia Naturalis, 9
Hitler, Adolf, 30
Holiday, where to go on, 8
Hollande, François, 58
Huntington-Hill method, 78, 86

I
Imperiali method, 80
Independence of irrelevant alternatives, 19,
    24, 27, 37, 38, 64
Independence of the powers, 3
Integrity, criterion of territorial, 109
Invisible hand, 124

J
Jefferson's method, 78, 80, 84
Jefferson, Thomas, 84
Judicial power, 3

K
Knox, Henry, 84

L
Laraki, Rida, 47
Largest remainder method, **73**, 84, 85, 100,
    101
Legislative power, 3
Le Pen, Marine, 58
Lippmann, Walter, 1, 12, 47
Lowndes, William, 85
Lull, Ramon, 11, 16, 20, 21

M
Majority
    absolute, 12

    criterion, 12, 41, 62, 64
    simple, 12, 29, 40, 70
Majority judgement, 6, 38, **47**, 92
    and Condorcet, 61
    collective ranking, 52
    domination, 55
    electoral ballot, 52
    majority gauge, 53
    majority grade, 49
    majority value, 50
Manipulability of vote, *see* vote, strategic
Marskin, Eric Stark, 6
Massachusetts, 105
Maximum consensus, 49
Mayor, election of, 30
May's theorem, 13, 41
Mean
    arithmetic, 23, 48, 78
    geometric, 78
    harmonic, 78
Median, 48
Mélenchon, Jean-Luc, 58
Mieux Voter, 67
Molise, 101, 102, 111
Monti, Mario, 6
Music contests, 10, 43, 65

N
Nash equilibrium, 67, 124
Nash, John, 124
Natural boundaries, criterion of, 109
Nazism, 30
Neutrality of collective choice functions, 41,
    42
New state paradox, 75, 85
No show paradox, 44, 64
Numbers, usage of, 23

O
Oklahoma paradox, 75
One person-one vote, 101, 102, 110, **113**
Ostrogorski paradox, 41

P
Parliamentary democracy, 4
Participation criterion, 44, 64
Partisan symmetry, 106
Pennsylvania, 106
Penrose index, 113
Plinius the eldest, 9
Plurinominal systems, 87

Political districting, **103**
   criteria, 109
Polsby-Popper metric, 111
Population paradox, 72, 75
Prisoner's dilemma, 124
Proportionality, degressive, 81

Q
Quota, 72, 94
Quota satisfaction, 72, 76, 96

R
Randolph, Edmund, 84
Range voting, 23, 26
Rank monotonicity, 45, 67
Rational preferences, 13, 36
Rounding, 72, 77, 78, 94, 97, 98, 100, 101
Run-off voting, **31**, 45

S
Sainte Laguë method, 80
Sainte Laguë modified method, 80
Sardinia, 101, 102, 110
Sarkozy, Nicolas, 58
Schulze
   method, 17
   table, 17
Seats
   cost, 71, 73, 82, 84, 85, 110
   monotonicity, 72
   plurinominal, 110
   population monotonicity, 72, 75
   representativity, 71
   uninominal, 30, 69, 103, 107, 110
Shakespeare, 25
Shapley-Shubik index, 4, 113
Shapley value, 113
Simeone, Bruno, 102, 103
Simple majority, 90
Skating, 8
Smith, Adam, 124
Social profile, 36
Social welfare function, 35
Sport competitions, **8**
Sport tournaments, 20

Strategic voting, 10, 23, 26, 42, 49, 64, 65, 67

T
Territorial units, 109
Tie-and-Transfer, 98
Tour de France, 9, 22
Trentino-Alto Adige, 101, 102, 110
Two alternatives, 41, 99

U
Unanimity, 37, 64
Uninominal systems, 87, 99
United Kingdom, 70
Universality of the domain, 36, 39, 40
UN Security Council, 114
US Congress, 74, 83
US Presidential election, 38, 118
US Supreme Court, 86, 106

V
Vinton, Samuel, 85
Von Neumann, John, 86
Vote
   anonymity, 88
   approval, 39, 54, 57, 92
   binary, 114
   distortion, 70, 99, 104
   instant run-off, 32, 91
   in the web, 67
   range, 23, 26
   run-off, 31, 45
   single, 11, 44, 57, 90
   single transferable, 91
   splitting, 30, 39, 90
   strategic, 10, 23, 26, 42, 49, 64, 65, 67
   ternary, 114
   weighted, 116
Voting power, **113**

W
Washington, George, 84
Webster, Daniel, 85
Webster's method, 78, 80, 85
Wines, 9

Printed in the United States
By Bookmasters